Polar and Climate Change Education

This book presents ideas for strengthening the foundations for transformational change in polar and global education leadership in all stages of the education process.

Despite being an established concept endorsed by the United Nations Educational, Scientific and Cultural Organization (UNESCO), Education for Sustainable Development (ESD) is still not in the educational mainstream but is vital in mitigating against the intensifying impacts of global change and adapting to the shifts that have already occurred. Drawing on examples from real world projects in the United States, Germany, Mexico, Japan, Peru and Greenland, this book assesses the new educational strategies, pedagogies and technologies which have been adopted by polar educators to stimulate students' interests in sustainability and re-orient education to global citizenship science. The experiential nature of the pedagogies shown in the case studies and educational activities builds background knowledge of cutting-edge research and empowers participants to communicate authentic research practices and show how data collection in the polar region is applicable in other parts of the globe.

Highlighting the many ways in which educators for global citizenship can have a decisive role in transforming individuals and society, this book will be of great interest to students and scholars of climate change, education and Arctic studies. It will also be a valuable resource for professional educators working in ESD.

Gisele Arruda is a professor in Circumpolar Studies (energy, Arctic, climate change, environment and society). She is a member of the University of the Arctic, the U.S. Arctic Research Committee, country director of International Higher Education Teaching and Learning Association, New York (USA) and a council member of Polar Educators International. She belongs to multidisciplinary research groups in Canada, the United States, Iceland, Norway, Greenland, conducting research on Arctic governance, sea-ice retreat, sustainable energy systems, smart specialization, climate change, societal studies, corporate social responsibility and multicultural aspects of higher education.

Research and Teaching in Environmental Studies

This series brings together international educators and researchers working from a variety of perspectives to explore and present best practice for research and teaching in environmental studies.

Given the urgency of environmental problems, our approach to the research and teaching of environmental studies is crucial. Reflecting on examples of success and failure within the field, this collection showcases authors from a diverse range of environmental disciplines including climate change, environmental communication and sustainable development. Lessons learned from interdisciplinary and transdisciplinary research are presented, as well as teaching and classroom methodology for specific countries and disciplines.

Interdisciplinary Research on Climate and Energy Decision Making
30 Years of Research on Global Change
Edited by M. Granger Morgan

Transformative Sustainability Education
Reimagining Our Future
Elizabeth A. Lange

Poetry and the Global Climate Crisis
Creative Educational Approaches to Complex Challenges
Edited by Amataritsero Ede, Sandra Lee Kleppe, and Angela Sorby

Research Journeys to Net Zero
Current and Future Leaders
Edited by Kyungeun Sung, Patrick Isherwood, and Richie Moalosi

Polar and Climate Change Education
Citizen Science and Sustainability
Edited by Gisele Arruda

For more information about this series, please visit: www.routledge.com/Research-and-Teaching-in-Environmental-Studies/book-series/RTES

Polar and Climate Change Education

Citizen Science and Sustainability

Edited by Gisele Arruda

Routledge
Taylor & Francis Group
LONDON AND NEW YORK

earthscan
from Routledge

First published 2025
by Routledge
4 Park Square, Milton Park, Abingdon, Oxon OX14 4RN

and by Routledge
605 Third Avenue, New York, NY 10158

Routledge is an imprint of the Taylor & Francis Group, an informa business

British Library Cataloguing-in-Publication Data
A catalogue record for this book is available from the British Library

ISBN: 978-1-032-78243-0 (hbk)
ISBN: 978-1-032-78244-7 (pbk)
ISBN: 978-1-003-48696-1 (ebk)

DOI: 10.4324/9781003486961

Typeset in Sabon
by Apex CoVantage, LLC

Contents

Contributors

Gisele Arruda is a professor in circumpolar studies (energy, Arctic, climate change, environment and society). She is a member of the University of the Arctic, the U.S. Arctic Research Committee, country director of International Higher Education Teaching and Learning Association, New York (USA) and a council member of Polar Educators International. She belongs to multidisciplinary research groups in Canada, the United States, Iceland, Norway, Greenland, conducting research on Arctic governance, sea-ice retreat, sustainable energy systems, blue economy, climate change, societal studies, corporate social responsibility and multicultural aspects of higher education. Her latest publications are *Corporate Social Responsibility in the Arctic* (Routledge) and *Climate Change Adaptation and Green Finance: The Arctic and Non-Arctic World*, in co-authorship with Lara Johannsdottir.

Inga Beck studied physical geography at the University of Munich and completed her PhD in 2011. Her main research focus was the detection of permafrost changes in the Canadian Arctic. Beside her scientific career, she is always very much interested in knowledge transfer and in bridging the gap between science and the public. Since the International Polar Year in 2007–2008, she was hence mainly involved in many education and outreach activities and was one of the first members of Polar Educators International. After she finished specialized scientific journalism studies in 2013 and a certificate as environmental pedagogue, she focused even more on education of sustainable development. She is the outreach officer of the Environmental Research Station Schneefernerhaus and Vice-President of Polar Educators International, while she is working self-employed as environmental educator for all kind of target groups.

Regina Brinker is an award-winning science and STEM (science, technology, education and mathematics) teacher. She believes that climate change is the most important issue facing students today, and works to engage students and communities with local and global environmental awareness and literacy. Regina has a strong interest in polar science. She participated in the PolarTREC project *Arctic Sunlight and Microbial Interactions* in 2014, spending three weeks in northern Alaska. Regina participated in the Fulbright-Japan Teacher Exchange for Education for Sustainable Development and the TOMODACHI STEM Teacher Academy. She has presented at American Geophysical Union meetings, 2010 International Polar Year (Oslo), Polar2018 (Davos), and National Science Teachers Association meetings. She wrote and narrated the TED-Ed talk *Phenology and Nature's Shifting Rhythms,* and participated in a Science Friday panel discussion *Bringing Climate Science into the Classroom.* Regina is now retired and lives with her husband in northern California.

Lars Demant-Poort is an associate professor for the Institute of Learning at Ilisimatusarfik, Greenland.

Gala Perez Gutierrez is Biologist from the Universidad Autonoma of Nuevo Leon, Master in Environmental Biology and Health from the Côte d'Azur University and Master in Scientific Information and Environmental Mediation from the Aix Marseille University. She has experienced as Environmental Educator and Scientific Outreacher in different institutions, universities and museums in Mexico and Europe. In the last few years, she has contributed to the actions for the Atoyac River restoration by raising awareness among citizens about the environment and carrying out clean-ups campaigns in France.

Louise Huffman is the Director of Education and Outreach at the U.S. Ice Drilling Program, Dartmouth College. Besides teaching children, Louise Huffman has been involved in teacher preparation, curriculum design, and teacher mentoring for more than 25 years. She is also on the leadership team of the National Science Foundation Science and Technology Center, COLDEX (Center for OLDest Ice EXploration) at Oregon State University (OSU – USA), including leading an educators' professional development workshop there each summer called the School of Ice. Through the years, her teaching career included students from ages 6–14 years old. As a teacher, she felt lucky to be selected for the National Science Foundation's Teachers Experiencing Antarctica Program and went to the Dry Valleys of Antarctica where she spent three months in a tent in the highest, driest, coldest place on Earth! Experience on that research team was incredibly exciting and challenging and the reason her life veered onto a path of polar education and climate change research communication. After retiring from teaching, Louise worked as the Coordinator of Education and Outreach for ANDRILL (ANtarctic geological DRILLing). The second time she went to Antarctica was with ANDRILL where she coordinated the on-ice experiences of an international team of eight teachers embedded on the ANDRILL field research team. An interesting note here is that the ANDRILL sediment cores are now housed at the OSU Marine Geology Repository where she is working with COLDEX.

Sebastian Krutkowski is an experienced teacher, published author and academic librarian with a passion for learning technologies and media. Having worked for more than ten years in the UK higher education sector, he is currently in his fourth year working at an international school in Tokyo. By practicing a strengths-based pedagogy, Sebastian strives to address the needs of diverse students in a culture of transition and mobility. His classes aim to develop responsible global citizens by focusing on a balanced cross-curricular approach which nurtures curiosity, problem solving and creativity, as well as commitment to equity, inclusion and social justice.

Esther Madrid Morales is Biochemical Engineer from the Instituto Tecnológico de Acapulco, Master in Environmental Sciences from the Universidad Autonoma del Estado de México, has experience in teaching at the High School level and is currently conducting doctoral research on the contamination of the Atoyac River, Guerrero, as part of her work as a PhD student in Environmental Sciences at the Universidad Autonoma de Guerrero.

Lionel Rogers, hailing from the Caribbean, is deeply committed to social equity and inclusivity, focusing on uplifting historically marginalized communities. As a cultural consultant, he promotes intercultural communication, understanding and appreciation. With expertise as an IDI qualified administrator, Lionel advocates fervently for mental and cultural health, working tirelessly to reshape educational landscapes for an ever-changing society.

Anne Farley Schoeffler has been teaching sixth, seventh and eighth grade science near Cleveland, Ohio, USA for 20 years and before that, taught Linguistics as an adjunct university professor in Fairfax, Virginia. She acted as a field research assistant in Greenland as part of PolarTREC ('Climate Change and Pollinators in the Arctic 2016'), serves on the board of the National Middle Level Science Teachers Association (former president), on the Council of Polar Educators International, and as a peer editor of the National Science Teaching Association's Science Scope magazine. She is passionate about environmental science and builds outdoor and exploratory investigations into her students' experiences.

Betty Trummel is a polar educator and an education outreach specialist from Massachusetts, USA.

1 Polar and Global Citizenship Education

How to Equip the Future Global Citizen Through Geo-Capabilities and Education for Sustainable Development

Gisele Arruda

Introduction

Sustainable development and education have an intrinsic, interdependent relationship with each other, largely because no sustainable society can operate without high-quality education and human development is intrinsically linked to education standards (AHDR-HH, 2014; Poppel, 2015, p. 67; Arruda, 2019). The current world context with its complexities, transitions and contradictions demand education beyond degrees and theory. It requires new ways of inculcating and stimulating live experiences on how to interact and engage with reality both under an individual and collective perspective, considering individual and societal needs. This creates the need for a necessary alignment between sustainable development and education in order to achieve tangible and practical results in terms of real-world solutions for contemporary issues based on responsibility, leadership and well-being.

If the main concern has been to expand the understanding of sustainability, making it more applicable and connected to global reality, new pedagogies, activities and praxis need to be developed and applied to education in order to incorporate different perspectives and context-specific factors (geomulticultural factors) of the reality much earlier than only higher education. This idea justifies the significance of developing geo-capabilities through curricula, as a specific branch of the capabilities approach to human development and economics. It is certainly a process that involves values to emerge from individuals to societies and from societies to the world in order to form the next generation of individuals who can deeply understand the concept and perspectives of sustainability and can efficiency exercise leadership and change. It also involves a new level of individual, collective and operational ethics that we need to debate and develop based on adaptative strategies for climate change.

Adaptive strategies are the ways in which individuals, households and communities change their productive activities and modify local rules and institutions to secure livelihoods (Berkes and Jolly, 2002, p. 18). The flexibility of

DOI: 10.4324/9781003486961-1

this more modern and resilient way of thinking and operating, in practice, allows communities to deal with climatic uncertainty and constantly adapt to change, in real time, by using information systems to disseminate their information within their communities. Adaptation depends on efficient communication that improves with a well-structured education and curriculum oriented to sustainable ways of development. The holistic dimension of Arctic issues requires an equally holistic approach to education in a way that information can be disseminated in a variety of formats intelligible to different stakeholders. Communication and education are powerful adaptation tools in a changing Arctic (Arruda, 2019).

The need for deeper and practical multidisciplinary understanding in relation to sustainable development and climate change is observed in different parts of the polar and non-polar world. In this chapter, the paradigm of Greenland's education system will be presented, and it will serve to highlight and expand the importance and the role of ESD, geo-capabilities and citizenship education to shape and equip the citizens and leaders of the present and the future.

ESD as a Powerful Adaptative Strategy

Adaptation seems to be the key word for the future under a multi-dimensional reality. It is a powerful word based on a challenging and impermanent reality of continuous complex interactions between the natural and social structures on Earth. These interactions are better understood when we study them through a geographical perspective (Arruda and Krutkowski, 2017), as some of these interactions, depending on the geographical area considered, can present dramatic consequences or create extraordinary geo-capabilities for learners. Considering that geography involves the holistic study of the interactions among human and natural systems (Norton and Mercier, 2016) and that the contemporary world presents a series of changes that will require severe physical and social adaptation, it is possible to argue that human geography and ESD education are more important than ever. The geophysical and socio-technical aspects of our world are naturally interlinked. With globalization, the human aspects of our existence are even more integrated, resulting in a range of collective opportunities, risks and responsibilities, as a civilization and as citizens.

The Arctic is the new frontier. It is a developing region in need of emergent and expanded understanding about the different complexities and dynamics of an extreme case study for climate change, socio-environmental adaptation and education. The search for natural and energy resources, the dynamism of entrepreneurship, new maritime routes for international trade and new extractive industrial activities in unexplored areas seem to be the trend for the Arctic region, triggering significant societal and developmental transformation. The challenge for polar education in this region is to understand these drivers of change and respond appropriately to them. Educational programs and scientific data have been carried out in Nordic countries, and education expertise clusters (expert institutes at Nordic universities) were studied to understand how sustainability and polar and climate change education are applied into teaching and learning in a more systematic way.

The speed of changes and eco insecurity makes it hard to predict the reality that global societies will face from 10–50 years into the future. Currently, it is not possible to determine with absolute precision all levels of vulnerabilities and multi-level impacts, or the scale of outcomes generated by combination of climate change, natural resources demand, pollution and human activity. It is also unknown how this set of risks will affect the ecosystem's ability to adapt to dynamic conditions. This level of complexity and unpredictability indicates the need to prepare students for a more sophisticated level of understanding, knowledge and skills to interpret and make decisions on sustainable development in a fast-paced, changing world. This is what is at stake in this chapter: the importance of effective preparation and adaptation for societies in times of fast-paced global environmental transition affecting the global commons.

Education is a powerful adaptation tool in a changing world. Education for Sustainable Development (ESD) plays a key role in local and global long-term societal adaptation strategies (Arruda, 2018a; Adger and Kelly, 1999). Adaptation depends on efficient communication that improves with a well-structured education and curriculum oriented to sustainable ways of development. The holistic dimension of global environmental issues requires an equally holistic and interdisciplinary approach to education in a way that knowledge and information can be disseminated in a variety of formats intelligible to different stakeholders from different geographic and cultural backgrounds. This need for improvement seems to occur due to a low curriculum adherence to the United Nations Educational, Scientific and Cultural Organization (UNESCO) ESD, a framework developed to promote "knowledge, skills, attitudes and values necessary to shape a sustainable future" (UNESCO, 2014b, p. 46) according to the concept of sustainable development. However, there seem to be additional factors contributing to this lack of clarity or engagement. The first one is the contradictions of the concept, the different interpretations and the fragmented nature of the educational systems during the last 30 years that confined this important discussion to silos. The Earth does not work in silos, as the human interaction affects several natural systems concomitantly, the approach related to sustainability needs to be comprehensive, interdisciplinary and systematic in several concomitant fronts.

ESD is not a new concept and emerged as a result of several international declarations, charters and programs in higher education from 1972–2014. Over the past 40 years of political activities in ESD, the institutional framework evolved from an orientation and experimental phase (1970–1990) to a transition and development phase (1990–2000) to an expansionary phase

(2000–2014), revealing a trajectory from an environmental perspective to a sustainability perspective (Michelsen, 2015). In fact, there have been many important educators who contributed to the development of ESD as it is today. Vare and Scott (2007) systematized levels of ESD in ESD 1 and ESD 2 (Vare and Scott, 2007, pp. 3, 4); Scott and Gough (2003, pp. 113, 116) proposed three types of ESD approaches, establishing a gradation of types (Type 1 – learning about SD; Type 2 – learning for SD; Type 3 – Learning as SD) (Scott and Gough, 2003, pp. 113, 116; Arruda, 2019). Another relevant model for ESD learning is the UNESCO's five pillars of learning (or five pillars of ESD) that was proposed by Jacques Delors et al. (1996a, 1996b) in their book *The Treasure Within* published by UNESCO. UNESCO's five pillars of learning is an approach to education based on lifelong learning and it consists of the following.

a) Learning to know is about having a broad general knowledge and in-depth understanding of a small number of subjects;
b) Learning to do is about having a main occupation but being skilled to deal with different situations and to work in teams;
c) Learning to live together is about understanding other people and our interdependence;
d) Learning to be is about personal development to make better choices and become more responsible;
e) Learning to transform oneself and society is about individuals working separately and together to change the world. This means gaining the knowledge, values and skills needed for transforming attitudes and lifestyles.

(Delors et al., 1996b, p. 37)

ESD, despite being an established concept endorsed by UNESCO with the purpose of integrating ESD in different educational sectors and with relevant initiatives like the United Nation's World Decade on Education for Sustainable Development (2005–2014), is still not the educational mainstream.

ESD is a vision of education that seeks to balance human and economic well-being with cultural traditions and respect for the earth's natural resources. ESD applies transdisciplinary educational methods and approaches to develop an ethic for lifelong learning; fosters respect for human needs that are compatible with sustainable use of natural resources and the needs of the planet; and nurtures a sense of global solidarity.

(UNESCO, 2005, p. 2) (Ryan, 2011, p. 3)

ESD is not the mainstream because these approaches and initiatives require further implementation and systematic application into practice in terms of educating all stakeholders about the interconnected global challenges under an acceptable perspective of sustainable development, geo-capabilities, SDGs and global citizenship precepts. ESD initiatives tended to focus on single projects to address sustainability in higher education, as opposed to taking a more systematic view of learning and change across the institutions to effectively address fundamental sustainability challenges at all levels of education (Mulà et al., 2017, p. 801; Tilbury et al., 2005, p. 1).

The concept of sustainability has focused on achieving human well-being and quality of life through the maintenance, care and equitable use of natural and cultural resources (Ryan, 2011, p. 3), but unfortunately, sustainable development has been used more as a catch-phrase than a revolution of thought, due to the great emphasis posed on economic growth. This unbalanced perspective undermined environmental reforms emphasized the resistance to change and the inherently contradictory ideology of SD. Hove (2004, p. 53) also argues that:

> Sustainable Development represents an altogether vague, inherently contradictory approach to mediating the impasse to development. Three main critiques were made: 1) sustainable development is Western construct, perpetuating the ideological underpinnings of former approaches, 2) it focuses its efforts on the unsustainable expansion of economic growth, and 3) its broad nature creates dangerous opportunities for actors to reinterpret and mould the approach the way they see fit.

The complexity around sustainability serves as an educational impulse for the improvement of learning processes, but ESD adherence and the metrics of these learning processes' outcomes require further development. The continuous search for sustainability presents an opportunity to develop learning activities and to explore debates over the issues at stake according to broader aims and approaches, experiential learning, critical thinking, reflection and critical pedagogies.

If the main aim of the ESD vision is to promote the balance between economic and human aspects of the Tripple Bottom Line (TBL), the ESD vision needs to promote, in practical terms, the fundamental sense of solidarity (compassion) that inspires the concept of global citizenship, as the latter elements of respect to the natural capacity and solidarity are not the mainstream in education – nor in civilization. In this sense, 'geographic' and 'cultural' are also pillars or tools for ESD toolbox proposed by this chapter in order to expand the understanding of the SD concept in a complex global context of polar and climate change education.

These perspectives indicate the importance of teaching skills or developing capabilities for complex thinking and future thinking, managing change by adopting a systems approach when covering the course content to stimulate an extended understanding or a holistic perspective of interconnected components that can integrate the economic, social, environmental, cultural and geographical interconnections. This is a challenge embraced by the education capability researchers as Young (2008) and Lambert (2014) whose positions are aligned to a 21st century knowledge based on "the counterintuitive sense of the planet as a place, with its physical and human interdependencies" (Lambert, 2014, p. 19).

Power Knowledge and Geo-Capabilities

The interdisciplinary knowledge base and geocultural perspective needed to understand the complexity of the polar and non-polar worlds under a systemic and holistic viewpoint seem to be supported by the components of what Young (2008, p. 14) refers to as "powerful knowledge":

> Powerful knowledge refers to what the knowledge can do or what intellectual power it gives to those who have access to it. Powerful knowledge provides more reliable explanations and new ways of thinking about the world and . . . can provide learners with a language for engaging in political, moral, and other kinds of debates.
>
> (Young, 2008, p. 14)

The reason of this knowledge to be powerful resides in the fact that "it provides the best understanding of the natural and social worlds" (Young, 2013, p. 196) by emphasizing the interconnections among elements, and it also enables people to envision alternatives (Young, 2014, p. 74) or futures scenarios. It opens possibilities, new ways of thinking about the world and consequently analyzing, understanding and participating in debates on significant local, national and global issues as it is systemic and specialized (Maude, 2015, p. 20). This is the knowledge that can enable students to become active citizens in the complex modern world (Young, 2011; Arendt and Jaspers, 1970). It has the power to reframe science and knowledge by paying attention to the local detail, or the geographical perspective of complexity, but at the same time to have on board the global perspective. Powerful knowledge expresses the link between complexity and geo-capabilities, providing a relevant framework to envision the large-scale processes of socio-environmental change which the polar and non-polar worlds face, and other realities that can put this concept in checkmate.

The different multicultural perspectives on the concept of SD have inspired researchers like me to progress the tools and the approaches used to educate stakeholders in a more inclusive and meaningful learning experience. More recent understandings of development process in different geographies emphasize the use and management of natural resources to satisfy human needs and improve people's quality of life. They also add concerns for people's health and education as key components to create more dynamic economies and higher material prosperity for societies.

Additionally, others view development as enabling people to live lives they value; however, it is important not to forget that values are different in different cultures.

One approach to development is the one in which the objective is to expand what people are able to do and be – or their real freedoms. This is a human-centered view of development whereby a healthy economy is one that enables people to enjoy a long and healthy life, a good education, a meaningful job, physical safety, democratic debate, and so on. According to this perspective aligned to Capability Theory, the analysis shifts from the economy to the person. The currency of assessment shifts from money to the things people can be and can do in their lives now and in the future. This is the view developed by the philosopher and Nobel laureate in economics Amartya Sen, whose writings on the 'capability approach' explain what capabilities are and provide the philosophical basis of human development. Under Sen's perspective, income is obviously an important instrument in enabling people to realize their full potential despite not being the most important. Consequently, the basic premise for development is to enlarge people's choices – and, consequently, their freedom (Alkire and Deneulin, 2009, p. 35). In this sense, the purpose of development is to enlarge all human choices, and the human development paradigm is concerned with building human capabilities and with using those human capabilities more fully (through an enabling framework for growth and employment). According to this perspective, human development should focus on four essential pillars: equality, sustainability, productivity and empowerment (Haq, 2004, p. 19). The economic aspect is obviously important, but this perspective emphasizes the need for quality and distribution for well-being and long-term sustainability. The human development paradigm defines the ends of development and analyzes sensible options for achieving them. This is the perspective we want to communicate and nurture in students of all levels of education.

The perspective of human development incorporates the need to remove – through their own capabilities – the obstacles that people face through the efforts and initiatives of people themselves. This effort toward collective well-being and freedom also improves societal organization and commitment. These are the two central ideas that give cogency to the focus on human development according to Amartya Sen's view (Fukuda-Parr and Kumar, 2003, p. 7).

The richness of this perspective points to the human development approach as being inherently multi-dimensional and plural. In practice, most policies focus on one or several components of human development; the approach itself is potentially broad. It is about education as much as it is about health. It is about culture as much as it is about political participation (Alkire and Deneulin, 2009, p. 28). The human development approach inspired by Amartya Sen's pioneering reflection in welfare economics and social and development economics has provided the basis of a new paradigm in economics and in the social sciences related to the capability approach. Amartya Sen writes:

> A person's capability to achieve 'functionings' that he or she has reason to value provides a general approach to the evaluation of social arrangements, and this yields a particular way of viewing the assessment of equality and inequality.
>
> (Sen, 1992, p. 5)

The key idea of the capability approach is that social arrangements should aim to expand people's capabilities – their freedom to promote or achieve what they value doing and being. An essential test of development is whether people have greater freedoms today than they did in the past. A test of inequality is whether people's capability sets are equal or unequal.

In terms of the capability approach, Sen (1999, p. 75) proceeds to define 'functionings' as "the various things a person may value doing or being" (Sen, 1999, p. 75). In other words, 'functionings' are valuable activities and states that make up people's well-being – such as being healthy and well-nourished, being safe, being educated, having a good job, being able to visit loved ones, and so on. They are also related to goods and income but describe what a person can do or be with these. For example, when people's basic need for food is met, they enjoy the functioning of being well-nourished. Capability refers to the freedom to enjoy various functionings. According to Sen's view, capability is defined as

> the various combinations of functionings (beings and doings) that the person can achieve. Capability is, thus, a set of vectors of functionings, reflecting the person's freedom to lead one type of life or another . . . to choose from possible livings.
>
> (Sen, 1992, p. 40)

Put differently, capabilities are "the substantive freedoms [a person] enjoys leading the kind of life he or she has reason to value" (Sen, 1999, p. 87). Finally, Sen defines 'agency' as the ability to pursue goals that one values and has reason to value. An agent is "someone who acts and brings about change" (Sen, 1999, p. 19).

The capability approach refers to the freedom to achieve well-being as a matter of what people are able to do and to be, and the kind of life they are effectively able to lead. In this sense, it is generally conceived as a flexible and multi-purpose framework, rather than as a precise theory of well-being (Sen, 1992, p. 48; Robeyns, 2005, pp. 94, 96). Capabilities are a person's real freedoms or opportunities to achieve functionings. Functionings are 'beings' and 'doings', or the various states of human beings and activities that a person can undertake. Functionings related to 'beings' are, for instance, being well-nourished, being undernourished, being housed, being educated, being illiterate, being literate, being healthy or being depressed. Examples of functionings related to 'doings' are taking part in a debate, voting in an election, traveling, killing animals, eating animals, consuming fuel in order to heat one's house and donating money to charity (Robeyns, 2016, p. 3). Functionings are constitutive of a person's being, and an evaluation of well-being has to take the form of an assessment of these constituent elements (Sen, 1992, p. 39). Sen continues:

> Whereas "functionings" are the proposed conceptualization for interpersonal comparisons of (achieved) well-being, "capabilities" are the conceptualization for interpersonal comparisons of the freedom to pursue well-being [which Sen calls "well-being freedom"].
>
> (Sen, 1992, p. 40)

Nussbaum (2000, p. 72) moves from theory to empirical applications when endorsing a specific list of capabilities. She justifies this list by arguing that:

> "Each of these capabilities is needed in order for a human life to be "not so impoverished that it is not worthy of the dignity of a human being. These capabilities are the moral entitlements of every human being on earth.
>
> (Nussbaum, 2000, p. 72)

In order to achieve their potential, people need to be able to stay healthy and take part in cultural, economic, social and political life. Broadly speaking they need to be in a position to take responsibility for their lives. They need to be able to think, make decisions and act according to what they believe is right. All these abilities, capacities, and attributes we refer to as human 'capabilities' (Nussbaum, 2006, p. 78). Applying a geographical approach to Nussbaum's capabilities, a framework on how geography can contribute to the development of people's intellectual functioning emerged as 'geo-capabilities' related to making choices for sustainability, being creative and productive in a global economy and culture, and achieving personal autonomy (Solem et al., 2013, p. 216):

> A capabilities approach to education asks teachers to consider the role of geography in helping young people reach their full human potential. Geography does not tell us how to live; but thinking geographically and developing our innate geographical imaginations can provide the intellectual means for visioning ourselves on planet earth.
>
> (Wadley, 2008, p. 650)

'Geo-capabilities' can be defined as the educational approach through which an individual can develop a greater potential to lead a life that they have reason to value, if they acquire geographical knowledge, enabling them to think geographically (Solem et al., 2013, p. 2). Geo-capabilities are focused on those capabilities in Nussbaum's list pertaining to human cognitive abilities and intellectual development, and then phrased in a manner that enables analysis of the curricular role of geography in helping young people think about their life in relation to themselves in the world and what may become of their communities, as well as people, places and environments around the world. The geo-capabilities project in education (initiated at school-level geography and applied to higher education) makes the explicit claim that the capabilities approach will enable and facilitate international communication about geography in education (Lambert et al., 2015, p. 217; Walkington et al., 2018, p. 11). They ensure the development of a progressive knowledge-led curriculum by seeking a different approach based on different perspectives. For curriculum design, this implies thinking about the role of geographic knowledge, skills, perspectives and values in developing the capabilities of young people. It also implies thinking in terms of how young people may become deprived of certain capabilities when they lack access to the powerful knowledge provided by geography education (Solem et al., 2013, p. 219; Donert, 2015, p. 2).

This perspective seems to be aligned to the ideas that curricula which only focus on competencies for paid employment are deficient (Marsh, 2009, p. 7), whereas capabilities represent more than only skills and knowledge. The developmental model of curriculum planning (Kelly, 2009, p. 99) suggests that educationists should consider particular views of humanity and human development, including social development and cultural components. In Kelly's words, "it sees the individual as an active being, who is entitled to control over his or her destiny and sees education as a process by which the degree of control available to each individual can be maximized"; or, in other words, "the central concern is the individual empowerment" (Kelly, 2009, p. 99). Consequently, the curriculum is more than stored knowledge, but it should be 'activity' or 'experience' (Board of Education, 1931, para. 75), a source of autonomy. Autonomy, according to Kelly (2009, p. 99) this means not only "freedom from constraints" but the ability to develop capacities that will enable the individual to make personal choices, decisions and judgments as an expression of a genuine control over his or her destiny. This notion applied to geographic knowledge, and the capacity of exercising this autonomy where one lives provides a powerful tool to live in the Arctic. This integral geographical perspective may equip students in higher education for the challenges of the 21st century, among them the challenges of a changing Arctic, being transformed by energy production and use (Boni and Walker, 2013, p. 20; Arruda, 2015, 2018a, 2018b).

This argument reinforces Ryan's idea (Ryan, 2011, p. 5) when arguing that a broad concept of 'holistic' curriculum change can be used to guide the review in its inclusive approach to these educational approaches that engage the entirety of the human personality and promote connectivity with the natural world. Practical and conceptual criteria can be established to set the boundaries and scope for the review, with both explicit and implicit links to ESD.

Interdisciplinary participatory pedagogies, 'real-world' research and a systemic view of sustainability are required to meet the transformative aim of education at all levels according to a citizen science approach. Sustainable development literacy, or eco-literacy, requires understanding of complex systems (Complexity Theory is employed to integrate human and physical aspects) (Dale and Newman, 2005, p. 354) and operating effects in cognitive (knowledge), affective (attitudes and values) and behavioral domains.

Global Citizenship Education (GCE)

SD requires not only urgent changes in energy systems, natural resources use and systems of production and consumption, but also new ways of viewing the world. To change the ways of viewing the world, it is necessary to develop new ways of thinking, acting and behaving. It implies new levels of responsibility informing values, skills and knowledge. It also implies not only the capacity of changing the world but reshaping ourselves to live in the world according to adaptive and flexible conduct.

Global citizenship means a sense of belonging to a broader community, beyond national boundaries, emphasizing our common humanity and the interconnectedness between peoples, as well as between the local and the global levels. It is based on the universal values of human rights, democracy, non-discrimination and diversity, and consists of voluntary practices oriented to social justice and global consciousness (UNESCO, 2016, p. 6). Global citizenship is fundamentally aligned to the following three specific ideas.

1. Learning to live together, self-identification with the whole of humanity, developing emotional intelligence (compassion), intercultural understanding to interact constructively across cultural boundaries (Haigh, 2014, p. 14).
2. Eco-literacy (learning to live together sustainably) to operate within the limits of the planet (Haigh, 2014, p. 14).
3. Responsibility, ethics, fairness and equity (Haigh, 2014, p. 14).

Self-identification with the whole of humanity refers to a sense of belonging to a broader community, beyond the boundaries of a national identity, a tribe or a nation, but referring to a planetary citizen, by understanding the interconnectedness and the interdependency of the beings living on planet Earth with a clear commitment to the collective good (UNESCO, 2014a, p. 14).

Eco-literacy is an educational paradigm expressing the ability to understand the natural systems through an integrated approach to deal with environmental problems related to the organization of ecosystems, energy systems and social systems (UNESCO, 2014a). Ecological intelligence allows comprehension of and ability to navigate the complexities between the natural and human systems' interaction (Goleman, 2009). The attribute of an ecologically literate person consists of learning to think the world in a holistic, complex, systemic and sustainable manner. Energy literacy is a fundamental aspect of eco-literacy because energy is in the genesis of the life processes.

Global citizenship implies a higher level of responsibility, ethics, fairness and equity in attitudes and decision-making (Arruda, 2019; Haigh, 2014, p. 14). Responsibility is embedded in global citizenship; it involves living responsibly by understanding that the effects of choices and actions taken in one place will certainly affect every level of life, family, community and the whole world (Arthur et al., 2008; Haigh, 2014, p. 16).

Theories of global citizenship involve at least three approaches related to political theory of global citizenship, educational theory of global citizenship and the social theory of global citizenship. The theoretical approach of this reflection is the educational theory of global citizenship or GCE that is represented by the components of the UNESCO Global Citizen Education Framework.

There is a powerful interface or relationship between GCE and ESD. GCE and ESD pursue the same vision of empowering students of all ages when providing means of understanding the interconnected elements of the world and the complexities of the global challenges faced by humanity in order to develop "global consciousness" and "global competence" (Dill, 2013). They both focus on global consciousness and competences involving the

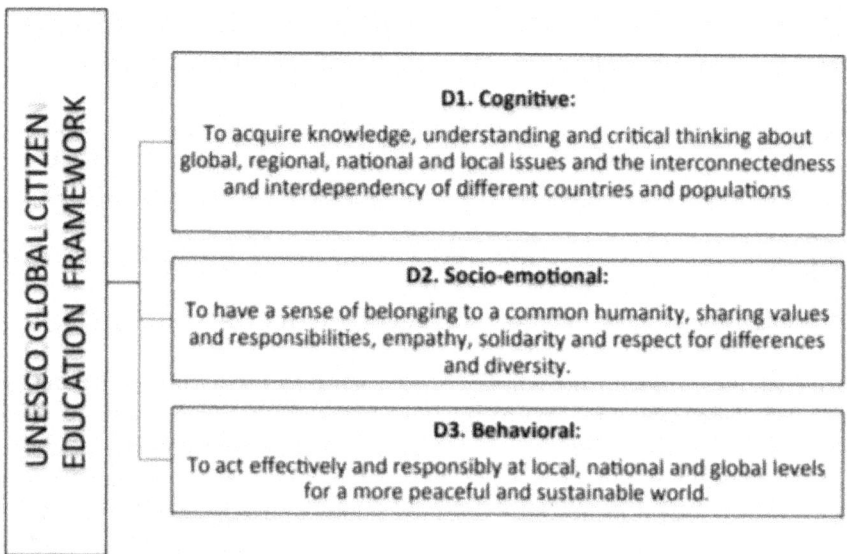

UNESCO GLOBAL CITIZEN EDUCATION FRAMEWORK

D1. Cognitive:
To acquire knowledge, understanding and critical thinking about global, regional, national and local issues and the interconnectedness and interdependency of different countries and populations

D2. Socio-emotional:
To have a sense of belonging to a common humanity, sharing values and responsibilities, empathy, solidarity and respect for differences and diversity.

D3. Behavioral:
To act effectively and responsibly at local, national and global levels for a more peaceful and sustainable world.

Figure 1.1 UNESCO Global Citizen Education Framework – core dimensions of global citizenship

Source: Pittman (2017)

Table 1.1 Global Citizenship Education: Topics and Learning Objectives

Learning Content	*Learning Outcome*	*Learning Process*
Action	Change	Transformation
Values	Behaviors	Addressing challenges
Collaboration	Communication	Critical thinking

Source: UNESCO (2015)

development of cognitive, socio-emotional and behavioral aspects that influence education in terms of contents, outcomes and learning processes. Values, change and transformation are at the center of GCE. The core dimensions of global citizenship involve the following.

These core conceptual dimensions serve as the basis for defining GCE goals, learning objectives and competencies, as well as priorities for assessing and evaluating learning. These core conceptual dimensions are based on, and include, aspects from all three domains of learning – cognitive, socio-emotional and behavioral – because global citizenship is not only about learning outcomes, but it is about competences and intelligences that go beyond the classroom activities.

The global citizen learner exercises the cognitive domain referring to thinking skills and the capacity of better understanding the world's complexities at the same time that this learner applies socio-emotional abilities related to values and social skills on how to live together with others respectfully and peacefully through the behavioral domain oriented to conduct, performance and engagement (UNESCO, 2015; Wolf et al., 2013).

These domains of learning derive from the pillars of learning from Delors et al. (1996b) in *Learning: The Treasure Within*:

1. Learning to know
2. Learning to do
3. Learning to be
4. Learning to live together
5. Learning to transform the world around

Considering the amplitude of components of global citizenship related to education, the implications for policy, curricula, teaching and learning are enormous (Albala-Bertrand, 1995; Banks, 2004).

> Global citizenship education applies a lifelong learning perspective, believing that knowledge is important to all ages at all domains of application, from early childhood through all levels of education and into adulthood, involving formal and informal approaches, curricular and extracurricular interventions, and conventional and unconventional pathways to participation.
>
> (UNESCO, 2014a, p. 46, 2015, p. 15)

Global citizenship education takes "a multifaceted approach, employing concepts and methodologies already applied in other areas, including human rights education, peace education, education for sustainable development and education for international understanding" and aims to advance their common objectives. Global citizenship education aims to be transformative, building the knowledge, skills, values and attitudes that learners need to be able to contribute to a more inclusive, just and peaceful world.

(UNESCO, 2014a, p. 14, 2015, p. 15)

While the global perspective is transformative and inclusive, it is worthwhile noting the potential of bioregions for addressing the sustainable development agenda. Bioregionalism or bioregional model is seen as an alternative strategy against the mechanistic thinking paradigm as defined by Aberley (1994) as a set of "knowledge and practices intending to reconnect societies in a sustainable way with their local and regional natural matrix". Its proposition is to have an integrative perspective of the local ecosystems, communities and local economic activities when arguing that real sustainability depends on recognizing the elements and characteristics of the bioregion because "every human community exists within a specific and unique bioregion consisting in the natural features that maintain life of that place", which has implications on bioregion planning and management (Toledo, 1999; Arruda, 2019, p. 98) and development of geo-capabilities. Stakeholders oriented to a sustainable vision should have the global perspective but should also have the ecological intelligence about the specific features and practices of different bioregions according to specific set of geo-capabilities. This global–local perspective explains important convergences aligned to new models, pedagogies and learning spaces to address change, future thinking and the complexities of climate and polar education.

The great contemporary challenge is to ensure that all learning institutions at all levels (from early childhood through adult education) embed ESD and SDGs, and that they transform their teaching and learning, infrastructure, operations, education governance, and engagement in order to provide sustainability and climate change education for social transformation. The education system as a whole also requires adaptation for green transformation, and it involves greening curriculum, schools, capacities, education systems and communities of practice. It also involves un-learning and re-learning the envisioning of a low-carbon economy and future.

ESD is well-positioned in the multilateral framework on climate change to progress climate change education and polar education by providing a lens to support goals, integrate climate action, provide clear criteria to policemakers and practitioners, and empower communities through regional and global strategies. ESD has been transforming learning environments through the application of new contents, pedagogies, values, skills and behaviors in order to align with sustainable development principles. It has the power of embedding sustainability into the education structure depending on how accurately one evaluates the adherence of courses, programs and curricula to ESD. This is an important aspect of ESD's practical implementation, considering the important role of education in leading society toward a more sustainable future by providing tangible educational contributions to SD (Cortese, 2003, 2012; Anderson and Herr, 1999) when assessing courses alignment (adherence) to ESD principles. In Arruda (2019), polar education studies based on circumpolar research helped create the Arctic Citizenship Model showing that learning should go beyond participation by proposing learning as an "active co-creator of adaptive solutions to complex and unpredictable real-world problems", a model that reveals that 'Education for Arctic Citizenship' goes beyond "learning as participation" (Vare, 2007, p. 4) – this is learning as sustainability in action. It is learning that is inculcated in the students' minds as per their practical (active) familiar, geo-experience of creativity and dissemination when living together. It is adaptive open-minded knowledge whose purpose is to effectively have impact in local, regional and global arenas. This is the 'education to be a geo-citizen', whereby the education site respects the 'space' and 'time' of the geographical area at stake – which can be the polar regions with their peculiar characteristics or the non-polar world and their intricate contexts. The Education for Arctic Citizenship co-relates to Type 4 and prepares students for Type 5, as shown in Table 1.2.

The Arctic citizen is eco-literate, energy literate and TK literate; his/her knowledge is active, dynamic and in constant adaptation, and it is used as an adaptation tool and empowerment for Indigenous and non-Indigenous groups, it is consequently not ethnocentric and focused on adaptability to complexity and dynamic systems accepting that SD is continuously in the making like the ESD model is in constant adaptation because knowledge is a dynamic system. Additionally, the polar research findings allowed Arruda (2019) to develop a set of components/parameters to evaluate the level of ESD practiced at the locations under study (Norway, Alaska, Canada) and proceed to the evaluation of the curricula adherence to ESD. The ESD Adherence Parameters Analytical Model proposed in Arruda (2019) allows educators and sustainability managers to compare the level of adherence of courses, programs and training by considering the presence and the absence of the components associated to the five types of ESD.

This reflection argues for the necessity to re-align all levels of education to specific competencies and intelligences necessary to proceed to a large-scale adaptation process that involves Indigenous and non-Indigenous peoples (Arruda and Krutkowski, 2016; Bitz et al., 2016) in the same scenario. This process would benefit if cognitive, affective, behavioral skills and knowledge would be connected to a range of specific capabilities privileging critical thinking, cultural competence for SD, complexity-thinking capabilities toward behavioral change and real-life practical knowledge application. The knowledge, skills and competences required to understand, live and operate in a low-carbon world require a more comprehensive range of competences

Table 1.2 Arctic Citizenship Model

Type 1 Learning About SD (Passive)	Type 2 Learning for SD (Action Is Desired)	Type 3 Learning as SD (Adaptive Learning)	Type 4 Education for Arctic Citizenship	Type 5 Education for Global Citizenship
Sustainable development as being expert-knowledge–driven whereby the role of the non-expert is to do as guided	Sustainable development as being expert-knowledge–driven whereby the role of the non-expert is to do as guided	Expert and learner work together for a tangible outcome (impact)	Multicultural experts and learners are co-creators and co-managers of knowledge Arctic Knowledge is the fruit of the encounter of documented types of knowledge (TK) and western knowledge Arctic Knowledge includes different perspectives (Indigenous and non-Indigenous perspectives) The Arctic citizen demonstrate geo-capabilities: • Geographical imagination • Ethical subject-hood • Integrative thinking about society and environment • Spatial thinking • A structured exploration of place The curriculum presents multicultural perspectives and foundations	Self-identified with the whole of humanity, not only tribal or group family Knowledge for living together sustainably, peacefully and responsibly

(Continued)

Assume that the problems humanity faces are essentially environmental (*it is just about management*)	Assume that our fundamental problems are social and/or political, and that these problems produce environmental symptoms and environmental impacts (*it is about economy, environment, society*)	Assume that what is (and can be) known in the present is not adequate	The Arctic citizen is eco-literate, energy literate and TK literate. The knowledge is active, dynamic and in constant adaptation, and it is used as an adaptation tool and empowerment for Indigenous and non-Indigenous groups – not ethnocentric (*adaptability to complex/dynamic systems; the SD is continuously in the making like the ESD model is in constant adaptation because knowledge is a dynamic system*)	The global citizen is eco-literate (energy literate) and interacts constructively across cultural boundaries
Can be understood through science and resolved by appropriate environmental and/or social actions and technologies (*the hope is technology to resolve problems*)	Such fundamental problems can be understood by means of anything from social-scientific analysis to an appeal to Indigenous knowledge	Desired 'end-states' cannot be specified	Arctic Knowledge is opened and constantly possible of being widened (*adaptability to unpredictability as knowledge is not static and can be widen by looking at the 'other' to different TKs; technology and social-technology; as systems are dynamic; learning is also dynamic; activity-active learning and application can create new realities*)	Knowledge oriented to the interconnectedness of different geographical areas; developing understanding, emotional intelligence (empathy and compassion)
It is assumed that learning leads to change once facts have been established and people are told what they are (*passivity*)	The solution in each case is to bring about social change (*activity and application are desired*)	This means that any learning must be open-ended	Knowledge creates solutions and new realities for the Arctic citizens that have a vision of the 'Arctic nation' and the Arctic citizen indistinctively of ethnicity (*knowledge to live together sustainably in the Arctic*)	Knowledge that embeds justice, fairness and responsibility

(Continued)

Table 1.2 (Continued)

Type 1 Learning About SD (Passive)	Type 2 Learning for SD (Action Is Desired)	Type 3 Learning as SD (Adaptive Learning)	Type 4 Education for Arctic Citizenship	Type 5 Education for Global Citizenship
	Learning is a tool to facilitate choice between alternative futures which can be specified on the basis of what is known in the present (learning can provide choices and decisions)	Essential if the uncertainties and complexities inherent in how we live now are to lead to reflective social learning about how we might live in the future.	No pedagogic imperialism; new way of being (authenticity and identity) being Indigenous or non-Indigenous, having the right to be and express yourself (individual and collective reflectivity- individual and collective critical thinking; critical reflection becomes a collective exercise; critical reflection is a social and multicultural process; critical reflection is a multicultural process; social critical thinking as learning in social action)

Leading to real-life learning experience

Creating solutions for real and complex problems
Making daily well-informed and responsible life decisions based on eco-literacy and energy literacy
Disseminating good practices
Affecting the area (local, regional, global) effectively
In the long term – intergenerational outcome and a future for all
For a sustainable way of life/durable model of existence and well-being for all | No pedagogic imperialism; a new way of being in the world – more inclusive

Creative co-existence of nature and humanity or planetary citizenship (Henderson and Ikeda, 2004) |

Source: Five types of ESD framework adapted from Scott and Gough (2003, pp. 113–116: Arruda, 2019, p. 175)

or intelligences aligned to a superior level of critical reflection oriented to long-term (intergenerational) collective outcomes based on multicultural respect and responsibility. By focusing on the levels of adherence to ESD, it is possible to enhance the chances of success in tacking issues of climate change, adaptive process and multicultural cooperation and co-management. The application of these models presented in this first chapter can also be part of global educational strategies to evaluate the courses alignment to ESD in other bioregions as a pro-active measure of conducting civilization to a more sustainable path of existence in the long run.

Another fundamental contribution resides in the fact that the models proposed in this chapter can be applied to evaluate the adherence of other initiatives to ESD to understand the main competences, intelligences and outcomes that educators and students in different geographical areas want to emphasize and to align to. Based on this combination of elements, new curricula, practices and pedagogies can be re-aligned, re-designed and re-adapted. The models can be applied in different geographical areas of the globe that are pursuing a change in the educational paradigm toward integrating ESD.

This transition needs to be accompanied by a powerful and culturally inclusive curriculum to provide the necessary practical knowledge and experience adapted to different ways of learning and able to forge a future sustainable low-carbon system that is fundamental for Arctic and non-Arctic nations, and so fundamental for polar and climate education to shape desired futures. Education oriented to develop an adaptive capacity for communities to respond to the scale of the current transition will be essential to secure the continuation, culture, and livelihoods of people; it is more than education for employability – it is intergenerational lifelong learning strategies.

Development that is durable means the ability to live a life according to a balanced functioning, well-being and freedom, according to Nussbaum (2000, p. 72), Sen (1999, p. 75), Freire (1998, p. 33) and Delors et al. (1996b, p. 37). It is not possible to exercise functionings, welfare and freedom without truly understanding the SD concept from a range of different perspectives and being able to apply them using a range of specific sustainability tools. Education oriented to values, context-specific factors (geo-multicultural competences), and with a vision toward geo-capabilities, contributes to citizens' capacity of making choices for sustainability. It contributes to shaping adequate policies, building capacities and mobilizing and empowering citizens, as well as accelerating action at local and global levels.

References

Aberley, D. (1994) *Futures by Design: The Practice of Ecological Planning*. New Society Publishers, Gabriola Island.

Adger, W.N. and Kelly, P.M. (1999) 'Social vulnerability to climate change and the architecture of entitlements'. *Mitigation Adaptation Strategies Global Change*, 4, 253–256, 254.

AHDR-HH. (2014) *Arctic Human Development Report II*. Norden, Copenhagen.

Albala-Bertrand, L. (1995) 'What education for what citizenship? First lessons from the research phase'. *Educational Innovation and Information, UNESCO IBE*, No. 82, Geneva.

Alkire, S. and Deneulin, S. (2009) 'The normative framework for development'. In Deneulin, S. with Shahani, L. (eds) *An Introduction to the Human Development and Capability Approach*. Earthscan, London, pp. 28, 35.

Anderson, G. and Herr, K. (1999) 'The new paradigm wars: Is there room for rigorous practitioner knowledge in schools and universities?'. *Educational Researcher*, 28(5), 12–21, 40.

Arendt, H. and Jaspers, K. (1970) 'Citizen of the world?'. In *Men in Dark Times*. Cape, London.

Arruda, G.M. (2015) 'Arctic governance regime: The last frontier for hydrocarbons exploitation'. *International Journal of Law and Management*, 57(5), 498–521, 500.

Arruda, G.M. (2018a) *Renewable Energy for the Arctic: New Perspectives*. Routledge, Abingdon.

Arruda, G.M. (2018b) 'Artic resource development. A sustainable prosperity project of co-management'. In Arruda, G.M. (ed) *Renewable Energy for the Arctic: New Perspectives*. Routledge, Abingdon, p. 112.

Arruda, G.M. (2019) *Sustainable Energy Education in the Arctic: The Role of Higher Education*. 1st edition. Routledge, Abingdon.

Arruda, G.M. and Johannsdottir, L. (2022) *Corporate Social Responsibility in the Arctic: The New Frontiers of Business, Management, and Enterprise*. Routledge, Abingdon.

Arruda, G.M. and Krutkowski, S. (2016) 'Arctic governance, indigenous knowledge, science and technology in times of climate change: Self-realization, recognition, representativeness'. *Journal of Enterprising Communities: People and Places in the Global Economy*, 11(4), 514–528, 519.

Arruda, G.M. and Krutkowski, S. (2017) 'Social impacts of climate change and resource development in the Arctic: Implications for Arctic governance'. *Journal of Enterprising Communities: People and Places in the Global Economy*, 11(2), 277–288, 279.

Arthur, J., Davies, I. and Hahn, C. (2008) *The SAGE Handbook of Education for Citizenship and Democracy*. SAGE Publications Ltd., London.

Banks, J. (2004) *Diversity and Citizenship Education: Global Perspectives*. John Wiley & Sons, San Francisco, CA.

Berkes, F. and Jolly, D. (2002) 'Adapting to climate change: Social-ecological resilience in a Canadian Western Arctic community'. *Conservation Ecology*, 5(2), 18.

Bitz, C., Blockley, E., Kauler, F., Petty, A., Massonet, F., Arruda, G.M., Sun, N. and Druckenmiller, M. (2016) 'Post-season report publishing'. Washington, DC. https://www.arcus.org/sipn/sea-ice-outlook/2016/post-season (accessed January 2017).

Board of Education. (1931) 'Primary education'. In *Hadow Report on Primary Education*. HMSO, London, para. 75.

Boni, A. and Walker, M. (2013) *Human Development and Capabilities: Re-imagining the University of the Twenty-First Century*. Routledge, New York, p. 20.

Cortese, A. (2003) 'The critical role of higher education in creating a sustainable future'. *Planning for Higher Education*, 31(3), 15–22.

Cortese, A. (2012) 'Foreword'. In Hemderson, H. (ed) *Becoming a Green Professional: A Guide to Careers in Sustainable Architecture, Development and More*. John Wiley, New York, 2012, pp. xi–xiii, 12.

Dale, A. and Newman, L. (2005) 'Sustainable development, education and literacy'. *International Journal of Sustainability in Higher Education*, 6(4), 351–362, 352, 354.

Davis, B. (2008) 'Complexity and education: Vital simultaneities'. In Mason, M. (ed) *Complexity Theory and the Philosophy of Education*. Wiley-Balckwell, Oxford, p. 47.

Delors, J. et al. (1996a) *L'éducation: Un trésor est caché dedans*. 1ére Edition. Éditions Odile Jacob, Paris.

Delors, J. et al. (1996b) *Learning: The Treasure Within: Report to UNESCO of the International Commission on Education for the Twenty-First Century*. UNESCO, Paris, p. 37.

Dill, J. (2013) *The Longings and Limits of Global Citizenship Education*. Routledge, New York.

Donert, K. (2015) 'GeoCapabilities: Empowering teachers and subject leaders'. The Innovative Pedagogies Series. https://www.heacademy.ac.uk/geocapabilities-empowering-teachers-subject-leaders, York, Higher Education Academy, p. 2 (accessed 12 December 2016).

Elkind, D. (2004) 'The problem with constructivism'. *The Educational Forum*, 68, 306–312, 307.

Elkind, D. (2005) 'Response to objectivism and education'. *The Educational Forum*, 69, 328–334.

Elkington, J. (1994) 'Towards the sustainable corporation: Win-win-win business strategies for sustainable development'. *California Management Review*, 36(2), 90–100, 92.

Elkington, J. (2004) 'Enter the triple bottom line'. In Henriques, A. and Richardson, J. (eds) *The Triple Bottom Line: Does It All Add Up? Assessing the Sustainability of Business and CSR Paperback*. Earthscan, London, pp. 23, 24.

Elkington, J. (2012) *The Zeronauts: Breaking the Sustainability Barrier*. Routledge, New York, p. 55.

Firth, R. (2011) 'Debates about knowledge and the curriculum: Some implications for geography education', In Butt, G. (ed) *Geography, Education and the Future*. Continuum, London, pp. 141–164.

Freire, P. (1998) *Pedagogy of Freedom: Ethics, Democracy and Civic Courage*. Rowan & Littlefield, Lanham, MD, p. 33.

Freire, P. (2000) *The Pedagogy of the Oppressed*. Alley Cat Editions, New York, NY, p. 50.

Fukuda-Parr, S. and Kumar, S.A. (2003) *Readings in Human Development*. Oxford University Press, Delhi, p. vii.

Fullan, M. (1989) 'Managing curriculum change'. In Preedy, M. (ed) *Approaches to Curriculum Management*. Open University Press, Milton Keynes, p. 144.

Gee, J.P. (2005) *An Introduction to Discourse Analysis: Theory and Method*. Routledge, New York, p. 97.

Geocapabilities. (2016) 'The Project'. http://www.geocapabilities.org/about/the-project/ (accessed 28 December 2016).

Gerring, J. (2007) *Case Study Research: Principles and Practices*. Cambridge University Press, New York.

Goleman, D. (2009) *Ecological Intelligence. How Knowing the Hidden Impacts of What We Buy Can Change Everything*. Broadway Books, New York.

Haigh, M. (2014) 'From internationalisation to education for global citizenship: A multi-layered history'. *Higher Education Quarterly*, 68(1), 6–27, 14, 16.

Haq, M. (2004) 'The human development paradigm'. In Fukuda-Parr and Kumar, S. (eds) *Readings in Human Development*. 2nd edition. Oxford University Press, New Delhi, pp. 17, 19.

Harman, W. (1988) *Global Mind Change: The Promises of the Last Years of the Twentieth Century*. Knowledge Systems Inc., Indianapolis, IN.

Henderson, H. and Ikeda, D. (2004) *Planetary Citizenship: Your Values, Beliefs, and Actions Can Shape a Sustainable World*. Middleway Press, Santa Monica.

Henriques, A. and Richardson, J. (2004) *The Triple Bottom Line, Does It All Add Up?* Earthscan Publications, London, pp. 23, 24.

Heron, J. (1996) *Cooperative Inquire. Research into the Human Condition*. Sage, London.

Hicks, D. (1991) 'Preparing for the millennium: Reflections on the need for futures education'. *Futures*, 23(6), 623–636.

Hill, J.D. (2000) *Becoming a Cosmopolitan: What It Means to Be a Human Being in the New Millennium*. Rowman and Littlefield, Lanham, MD.

Hinchliffe, G. and Terzi, L. (2009) 'Introduction to the special issue "Capabilities in education"'. *Studies in Philosophy and Education*, 28, 387–390.

Hove, H. (2004) 'Critiquing sustainable development: A meaningful way of mediating the development impasse?'. *Undercurrent*, I(1), 48–54, 53.

Hovelsrud, G. and Smit, B. (2010) *Community Adaptation and Vulnerability in Arctic Regions*. Springer, Dordrecht, New York, pp. 3, 69, 71, 75.

Hunter, J.L. and Krantz, S. (2010) 'Constructivism in cultural competence education'. *Journal of Nursing Education*, 49(4), 207–114, 207, April.

Hutchins, R. (1968) *The Learning Society*. University of Chicago Press, Chicago, IL, p. 1.

Intergovernmental Panel on Climate Change. (2000) *Emissions Scenarios. Special Report of the Intergovernmental Panel on Climate Change*. Marinovic, N. and Swart, R. (eds). Cambridge University Press, Cambridge.

James, W. (1977) *A Pluralistic Universe*. Harvard University Press, Cambridge, MA (Original work published 1909), p. 117.

Kelly, A.V. (2009) *The Curriculum. Theory and Practice*. 6th edition. Sage, London, pp. 88, 99.

Kvale, S. and Brinkman, S. (2009) *Interviews: Learning the Craft of Qualitative Research Interviewing*. 2nd edition, Sage Publications, Inc., Thousand Oaks, CA.

Lambert, D. (2014) 'Curriculum thinking, "capabilities" and the place of geographical knowledge in schools'. *Prace Komisji Edukacji Geograficznej*, t. 3, s. 13–30, p. 19.

Lambert, D., Solem, M. and Tani, S. (2015) 'Achieving human potential through geography education, a capabilities approach to curriculum making in schools'. *Annals of the Association of American Geographers*, 217. http://dx.doi.org/10.10 80/00045608.2015.1022128.

Larsen, J.N. and Fondahl, G. (2015) *Arctic Human Development Report: Regional Processes and Global Linkages*. Nordisk Ministerråd, Copenhagen, p. 500.

Longman, M. (2014) 'Aboriginography. A new decolonized aboriginal methodology'. In Guttorm, G. and Somby, S.R. (eds) *Diedut, V.3*. Sami University College, Alta, pp. 16, 17.

Marsh, C.J. (2009) *Key Concepts for Understanding Curriculum*. 4th edition. Routledge, Oxford, p. 7.

Mason, M. (2008a) *Complexity Theory and the Philosophy of Education*. Wiley-Blackwell, Oxford, pp. 12, 33, 216.

Mason, M. (2008b) 'What is complexity theory and what are its implications for educational change?'. In Mason, M. (ed) *Complexity Theory and the Philosophy of Education*. Wiley-Balckwell, Oxford, p. 45.

Maude, A. (2015) 'What is powerful knowledge, and can it be found in the Australian geography curriculum?'. *Flinders University, Adelaide, South Australia Geographical Education*, 28, 18–26, 20.

Mayr, E. (1991) *One Long Argument: Charles Darwin and the Genesis of Modern Evolutionary Thought*. Harvard University Press, Cambridge, MA, p. 97.

Michelsen, G. (2014) 'Education for sustainable development. Status quo and perspectives'. In O'Farrel, L., Schonman, S. and Wagner, E. (eds) *Yearbook of Research in Arts Education*. Waxmann, Münster, pp. 121–129.

Michelsen, G. (2015) 'Policy, politics and polity in higher education for sustainable development'. In Barth, M., Michelsen, G., Rieckmann, M. and Thomas, I. (eds) *Routledge Handbook of Higher Education for Sustainable Development*. Routledge, Abingdon.

Morcol, G. (2001) 'Positivist beliefs among policy professionals: An empirical investigation'. *Policy Sciences*, 34, 381–401.

Morrison, K. (2008) 'Educational philosophy and the challenge of complexity theory'. In Mason, M. (ed) *Complexity Theory and the Philosophy of Education*. Wiley-Balckwell, Oxford, pp. 16, 29, 45.

Mulà, I., Tilbury, D., Ryan, A., Mader, M., Dlouhá, J., Mader, C., Benayas, J., Dlouhý, J. and Alba, D. (2017) 'Catalysing change in higher education for sustainable development: A review of professional development initiatives for university educators'. *International Journal of Sustainability in Higher Education*, 18(5), 798–820, 801.

Norton, W. and Mercier, M. (2016) *Human Geography. OUP Catalogue*. 9th edition. Oxford University Press, Oxford.

Nussbaum, M. (2000) *Women and Human Development: The Capabilities Approach*. Harvard University Press, Cambridge, MA, pp. 70–77, p. 72, 73, 74.

Nussbaum, M. (2006) *Frontiers of Justice: Disability, Nationality, Species Membership*. Harvard University Press, Cambridge, MA, pp. 78–81, 78.

Nussbaum, M. (2011) *Creating Capabilities: The Human Development Approach*. Harvard University Press, Cambridge, MA, pp. 33–34, 33.

Nussbaum, M. and Sen, A. (1993) *The Quality of Life*. Clarendon Press, Oxford University Press, Oxford, England, New York, p. 20.

Orr, D.W. (1992) *Ecological Literacy: Education and the Transition to a Postmodern World*. SUNY Press, Albany, 210 p.

Palmer, J. (2003) *Environmental Education in the 21st Century: Theory, Practice, Progress and Promise*. Routledge, London, pp. 275, 276.

Peters, M.A., Britton A. and Blee, H. (2008) *Global Citizenship Education: Philosophy, Theory and Pedagogy*. Sense Publishers, Rotterdam.

Pieterse, J.N. (2010) *Development Theory. Deconstructions/Reconstructions*. 2nd edition, Sage, London, p. 3.

Pittman, J. (2017) 'Exploring global citizenship theories to advance educational, social, economic and environmental justice'. *Journal of Tourism and Hospitality*, 6, 326.

Poppel, B. (2015) *SliCA: Arctic Living Conditions – Living Conditions and Quality of Life among Inuit, Sami and Indigenous Peoples of Chukotka and the Kola Peninsula*. Nordic Council of Ministers, Denmark, p. 67.

Roberts, M. (2014) 'Powerful knowledge and geographical education'. *The Curriculum Journal*, 25, 187–209.

Robertson, M. (2014) *Sustainability Principles and Practice*. 1st edition. Routledge, Oxford, p. 7.

Robeyns, I. (2005) 'The capability approach: A theoretical survey'. *Journal of Human Development*, 6, 93–114, 94, 96.

Robeyns, I. (2016) 'The capability approach'. In Zalta, E.N. (ed) *The Stanford Encyclopedia of Philosophy*. Winter 2016 edition, p. 3. https://plato.stanford.edu/archives/win2016/entries/capability-approach/ (accessed 13 January 2016).

Ryan, A. (2011) *ESD and Holistic Curriculum Change*. http://www.heacademy.ac.uk/resources/detail/sustainability/esd_ryan_holistic, pp. 3, 5. (accessed 12 December 2016).

Scott, D. (2005) 'Critical realism and empirical research methods in education'. *Journal of Philosophy of Education*, 39(4), 633–646, 634.

Scott, W.A.H. and Gough, S.R. (2003) *Sustainable Development and Learning: Framing the Issues*. Routledge Falmer, London, pp. 113, 116.

Sen, A. (1990) 'Development as capability expansion'. In Griffin, K. and Knight, J. (eds) *Human Development and the International Development Strategy for the 1990s*. Macmillan, London, pp. 41–58.

Sen, A. (1992) *Inequality Re-Examined*. Clarendon Press, Oxford, pp. 5, 39, 40, 48.

Sen, A. (1993) 'Capability and well-being'. In Nussbaum, M. and Sen, A. (eds) *The Quality of Life*. Clarendon Press, Oxford, pp. 30–53, 31.

Sen, A. (1999) *Development as Freedom*. Oxford University Press, Oxford, pp. 19, 75, 87.

Sen, A. (2004) 'UN human development report 2004: Chapter 1 Cultural liberty and human development'. In *UN Human Development Reports*. United Nations Development Programmer, p. 16, http://hdr.undp.org/sites/default/files/reports/265/hdr_2004_complete.pdf (accessed 10 December 2016).

Sen, A. (2009) *The Idea of Justice*. Harvard University Press, Cambridge, MA.

Senge, P. (1990) *The Fifth Discipline: The Art and Practice of the Learning Organization*. Century Business, London, p. 68.

Smit, B. and Wandel, J. (2006) 'Adaptation, adaptive capacity, and vulnerability', *Global Environmental Change*, 16, 282–292, 282, 283.

Smith, L.T. (1999) *Decolonizing Methodologies: Research and Indigenous Peoples*. Zed Books, New York.

Smuts, J. (1999) *Holism and Evolution*. Sierra Sunrise Publishing, Sherman Oaks, CA, p. 87.

Solem, M., Lambert, D. and Tani. (2013) 'GeoCapabilities: Toward an international framework for researching the purposes and values of geography education'. *Review of International Geographical Education Online*, 3(3), 204–219, 216, 219.

Stengers, I. (1997) *Power and Invention: Situating Science*. University of Minnesota Press, MN.

Stephens, S. (2000) *Handbook for Culturally Responsive Science Curriculum*. Alaska Science Consortium and the Alaska Rural Systemic Initiative, Fairbanks, AK, p. 10.

Sterling, S. (2003) 'Whole Systems Thinking as a basis for paradigm change in education: Explorations in the context of sustainability'. Unpublished doctoral dissertation, University of Bath, Bath.

Tilbury, D., Keogh, A., Leighton, A. and Kent, J. (2005) 'A national review of environmental education and its contribution to sustainability in Australia: Further and higher education'. Report prepared by Australian Research Institute in Education for Sustainability (ARIES) for the Department of the Environment and Heritage, Australian Government, Sydney, p. 1. http://www.aries.mq.edu.au/project.htm (accessed 30 November 2018).

Toledo, V. (1999) Consensos Naturo-Sociales: Una Evaluación de las Nuevas Construcciones del Territorio y de las Regiones, Comité Técnico Inter-agencial del Foro de Ministros de Medio Ambiente de América Latina y el Caribe, doc. Mimeo, julio, 1999.

UN General Assembly. (2015) *Transforming Our World: The 2030 Agenda for Sustainable Development*, 21 October 2015, A/RES/70/1, p. 14. https://www.refworld.org/docid/57b6e3e44.html (accessed 5 January 2019).

UNESCO. (2005) *United Nations Decade of Education for Sustainable Development (DESD) 2005–2014*. UNESCO, Paris, p. 2.

UNESCO. (2014a) *Education Strategy 2014–2021*. UNESCO, Paris, p. 46. http://www.natcom.gov.jo/sites/default/files/231288e.pdf (accessed 2 December 2018).

UNESCO. (2014b) *Global Citizenship Education. Preparing Learners for the Challenges of the 21st Century*. UNESCO, Paris, p. 14.

UNESCO. (2015) *Global Citizenship Education: Topics and Learning Objectives*. UNESCO, Paris, p. 15. https://unesdoc.unesco.org/ark:/48223/pf0000232993 (accessed 26 November 2018).

UNESCO. (2016) *Schools in Action. Global Citizens for Sustainable Development. A Guide for Students*. UNESCO, Paris, p. 6. https://www.unesco-sole.si/doc/teme/sdg-guide-for-teachers.pdf (accessed 25 November 2018).

Vare, P. (2007) 'From practice to theory: Participation as learning in the context of sustainable development projects'. In Reid, A.D., Jensen, B.B., Nikel, J. and Simovska, V. (eds) *Participation and Learning: Perspectives on Education and the Environment, Health and Sustainability*. Springer Press, Dordrecht.

Vare, P. and Scott, W.A.H. (2007) 'Learning for a change: Exploring the relationship between education and sustainable development'. *Journal of Education for Sustainable Development*, 1(2), 3, 4.

Wadley, D. (2008) 'The garden of peace'. *Annals of the Association of American Geographers*, 98(3), 650.

Walkington, H., Dyer, S., Solem, M., Haigh, M. and Waddington, S. (2018) 'A capabilities approach to higher education: Geocapabilities and implications for geography curricula'. *Journal of Geography in Higher Education*, 42(1), 7–24, 11.

Wolf, J., Allice, I. and Bell, T. (2013) 'Values, climate change, and implications for adaptation: Evidence from two communities in Labrador, Canada'. *Global Environmental Change*, 23(2), 548–562, 550.

Young, M. (2008) *Bringing Knowledge Back In: From Social Constructivism to Social Realism in the Sociology of Education*. Routledge, Abingdon.

Young, M. (2011) 'The future of education in a knowledge society: The radical case for a subject-based curriculum', *Journal of the Pacific Circle Consortium for Education*, 22(1), 21–32, December.

Young, M. (2013) 'Powerful knowledge: An analytically useful concept or just a 'sexy sounding term? A response to John Beck's 'Powerful knowledge, esoteric knowledge, curriculum knowledge', *Cambridge Journal of Education*, 43, 95–198, 196.

Young, M. (2014) 'Powerful knowledge as a curriculum principle'. In Young, M., Lambert, D., Roberts, C. and Roberts, M. (eds) *Knowledge and the Future School: Curriculum and Social Justice*. Bloomsbury Academic, London, pp. 65–88, 74.

2 The School of Snow and Ice (SNOWI)

Professional Development for Middle School and High School Educators in Ice Core Science and Climate Change

Louise Huffman and Lars Demant-Poort

Introduction

This chapter describes the School of Snow and Ice (SNOWI), a new two-part professional development program for educators of middle school and high school students (ages 11–18). SNOWI trains educators to understand ice core and climate change science and to acquire the skills necessary to bring this exciting inquiry into their classrooms. The experiential nature of this workshop builds teachers' background knowledge of cutting-edge research and empowers them to communicate authentic research practices, ice core data and results to their students. Embedded in the workshop are hands-on labs based on inquiry teaching (Pedaste et al. 2015; Chu et al. 2017) and constructivist learning skills (Jones et al. 2002) that can be used in almost any curriculum or subject area. SNOWI is a program that has grown out of the School of Ice (SOI) as a rigorous professional development program of the U.S. Ice Drilling Program (IDP), funded by the National Science Foundation (NSF) for faculty at two- and four-year Minority Serving Institutions. SOI has been providing cutting-edge science and resources since 2015. SNOWI and the Geo-Bridge program for high school teachers and community college faculty have evolved from SOI to fill a need to engage students in climate change science and sustainability education earlier than college-level courses. SNOWI was piloted by IDP in partnership with the University of Greenland and was funded by the NSF, the Danish Embassy and the University of Greenland. The Geo-Bridge program will have its inaugural debut summer of 2024, also funded by NSF.

The purpose of this chapter is to share lessons learned from conducting an international workshop for educators that may spark ideas with anyone in a position of creating professional development experiences for educators. Maybe even more importantly, we share here a treasure trove of ice core and climate change labs, activities and demonstrations that made this workshop a success and are now being used in many teachers' classrooms.

The SNOWI Workshop was designed for teachers of students ages 11–18 and is based on the SOI program model (https://icedrill-education.org).

DOI: 10.4324/9781003486961-2

SNOWI's original plan was a two-part workshop including specific activities for teachers to integrate into their classrooms. Workshop 1 introduced inquiry teaching and constructivist learning through hands-on science activities and snow and ice labs. After Workshop 1, pre-service teachers presented one or two labs to students during their student teaching experience. In-service teachers were also expected to incorporate at least one lab into the curriculum with their students. Workshop 2 built on the knowledge base from the first workshop and brought cutting-edge research, lab experiences, scientist presenters and field experiences designed to increase teachers' background knowledge, as well as provide teaching resources to transfer that knowledge to their students.

Teachers wore two "hats" during the workshops. During learning activities, they were in the role of students as they worked through the hands-on labs, but during "teacher moments," the modeling of teaching was highlighted and discussed, including specific inquiry teaching and constructivist learning techniques and skills. The workshops took place at Ilisimatusarfik, the University of Greenland in Nuuk. The original plan was for two IDP educators to facilitate an in-person workshop including both **pre-service teachers** in the teacher training program at the university and **in-service teachers** from settlements and towns in Greenland. The professor in charge of the teacher training module was to be a co-facilitator during the workshop, but COVID-19 caused many changes to the original plans for this program. One of the main effects of COVID-19 was the impossibility for the U.S. facilitators to travel to Greenland to present in person, so a hybrid model was created. The U.S. facilitators were the creators of the resources used in the workshop, so they were deeply and thoroughly familiar with them. They led the activities from the "Zoom Wall," while the professor oversaw organizing all of the materials ahead of time, monitoring learning during the activity and clean-up after.

One of the hardest things as a virtual workshop facilitator is the difficulty of "reading" the audience, something that happens naturally in person. It is important to be able to read body language and respond on the spot to meet participants' needs and to recognize when you need to slow down, speed up, or re-teach your presentation. The professor was able to "read" the room for the virtual presenters, letting them know when it was time to stop for questions and/or additional directions. To complicate the process more, there were three primary languages in the workshop: English, Danish and Greenlandic. To see how languages were handled see the section on "Challenges."

Workshop 1: Introducing Inquiry Teaching and Constructivist Learning

On each day of the workshop, a "Live Agenda" was built by linking each individual activity to the agenda as soon as it was completed by the participants.

At the end of each day during the workshop series, the Live Agenda provided teachers immediate access to that day's activities, labs, presentations, and the workshop recordings for future reference. For the purposes of this article, an abbreviated version of the Live Agenda provides you with links to all hands-on activities mentioned here, and much more (https://docs.google. com/document/d/170rvx5T7EgIJqM5pGNtxKosDOHKmWO9yJUz30IgA Yvk). You might find it helpful to keep the Live Agenda open to refer to while you read this article. We hope you find this a rich resource for your own educational purposes.

In Workshop 1 of SNOWI, participants were introduced to inquiry through hands-on science labs explicitly designed to focus on specific aspects and techniques of inquiry teaching. In this section of the workshop, teachers were introduced to inquiry teaching (what the teacher does) and constructivism (what the students do) through experiential activities created by the facilitators, but based on best practices for educators. For example, an introductory activity was "Rocket Blast-Off," followed by the "Payload Challenge." The important differences between a fun activity and an inquiry learning experience were highlighted in the comparison of these two activities.

Participants were introduced to the "Eternal Verities" about teaching (Bodner 1986). These "truths" are timeless and cause early career educators to look quizzical, and seasoned teachers to smile and nod their heads. The Eternal Verities were presented through personal stories and hands-on demonstrations. Some examples are:

- Teaching and learning are not synonymous.
- We can teach, and teach well, without having students learn.
- Knowledge is seldom transferred intact from the mind of the teacher to the mind of the student.
- When placed in a stimulating environment with enthusiastic people, some who think they do not want to learn change their minds.
- Active students learn more than passive ones.
- No one in the classroom learns more than the individual teaching the course for the first time.

Another important aspect of the workshop series is the use of science notebooks for learning. The contents of a student notebook can provide a window into how your students think, and the outside of the notebook is a window into who your students are. (See "Synergistic Notebooks" in the Live Agenda) The Next Generation Quick Reference Guide to the NGSS, p. 25 (Willard 2015), emphasizes the importance of notebooks:

Students should write accounts of their work, using journals to record observations, thoughts, ideas, and models. They should be encouraged to create diagrams and to represent data and observations with plots and tables, as well as with written text, in these journals.

Science Notebooks played a critical role in the workshops. As labs and activities were presented, participants started a new two-page spread. The first page was used to record details of the lesson, along with sketches, graphs and directions. On the second page, they recorded their thoughts about how they might use the activity in their classrooms, including any modifications they would make to differentiate it for their students. At the end of the workshop, educators had a thorough record of explorations they experienced, and they gained an understanding of how this tool can be a powerful learning experience for their students, as well.

Changing science teaching from a teacher-focused classroom to a student-centered one is not easy. One tool that can help with this is the Wheel of Inquiry created by Wendy Pierce (2001). This tool helps students identify variables and investigable questions for planning and conducting their own scientific inquiry. These explorations then lead students to the construction of reasonable explanations and communication of the investigation.

During Workshop 1, we introduced a simplified "Learning Cycle" originally created by David Walbert (2014): exploration, concept development, concept application. This cycle incorporates an important piece of

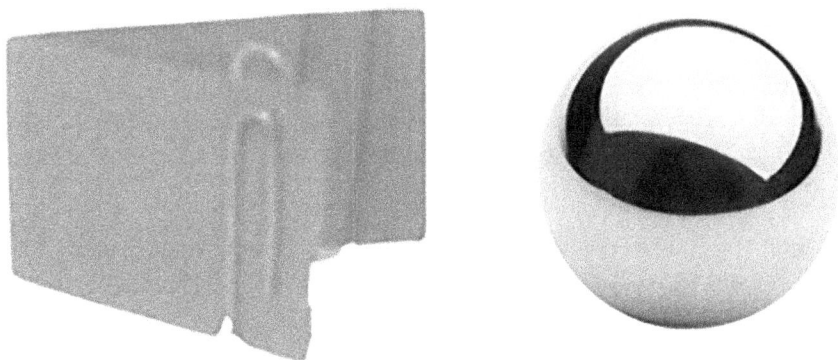

Figure 2.1 "Energy sphere" and index card "snowplow"

Figure 2.2 Ruler ramp set-up

constructivist learning: activating prior knowledge through the ABCs of learning: *Activity Before Concepts*.

1. During exploration, students were introduced to the materials and system to be used. In our "Snowplow Sliders" example (see Live Agenda) students were given an index card, paper clips, a marble, a ruler-ramp and blocks to elevate the ramp. They were given time to explore how the system worked and were challenged to make a "snowplow" that could be pushed by a marble ("energy sphere") without tipping over. (See slide No. 20 in "Snowplow Sliders" in Live Agenda).
2. During Concept Development, students were given a question to explore. Using their notebooks, they conducted their experiments and created data tables and graphs.
3. In the Concept Application stage, students identified dependent and independent variables and created an investigable question of their own and then carried out the investigation.

On Day 2 of Workshop 1, we conducted activities to distinguish between science and engineering. The IDP is a real-world example of how scientists and engineers must work together to conduct ice core drilling and science research projects. IDP helps coordinate the U.S. ice drilling community to develop a 10-year "Long Range Science Plan" to articulate goals and make recommendations for the direction for U.S. ice coring and drilling science in a wide variety of areas of scientific inquiry (Ice Drilling Program 2023b). This science plan makes recommendations for the development of drilling

	Science	*Engineering*
Focus	Science begins with a **question about a phenomena**, and seeks to find theories that can provide explanatory answers **(inquiry)**	Engineering begins with a **problem, human need or desire** that suggests an engineering problem that needs a **solution (design)**
Outcomes	New knowledge	New products or solutions

Figure 2.3 The similarities and differences between science and engineering

technology, infrastructure and logistical support needed to enable the science. A companion document, the "Long Range Drilling Technology Plan," provides details about drill capabilities currently available and those needing to be developed (Ice Drilling Program 2023a).

For teachers, we introduce engineering projects like the "Long Haul" which highlight the similarities and differences between science and engineering.

The "Egg Drop" activity introduces the "Engineering Design Cycle.

Similar to the Learning Cycle, the Engineering Design Cycle moved from an idea about how to solve a problem to designing a solution, building it, testing it and improving it. The important part for teaching engineering is to give time to re-design and re-enter the cycle until a final improved product is completed.

Teachers were reminded of Bloom's Taxonomy and the importance of asking students higher-level questions (Anderson et al. 2001). As a group, we asked the teachers to think of a question they might ask during a science activity they teach in their curriculum. In collaborative groups, they were challenged to rewrite those questions at each level of Bloom's Taxonomy.

- Remember
- Understand
- Apply
- Analyze
- Evaluate
- Create

Many of us learned about Bloom's Taxonomy in college, but to internalize it and comfortably generate thought-provoking questions in class requires much practice and fine-tuning of that skill over time.

In preparation for presenting labs and activities to their students, teachers participated in lesson development based on *Understanding by Design*

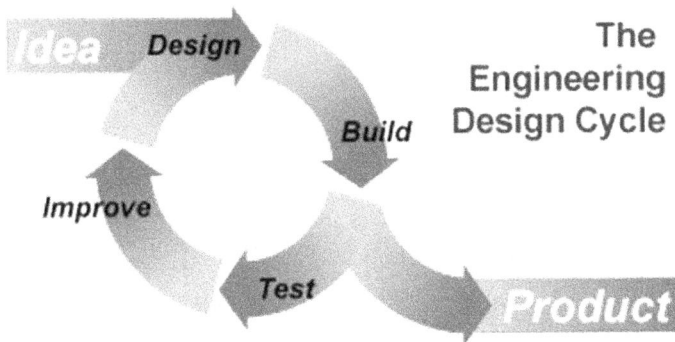

Figure 2.4 Engineering design cycle

(Wiggins and McTighe 2005; Bowen 2017). A distinction was made between planning a lesson (backwards design) and teaching a lesson.

Workshop activities for the participants were designed to be engaging, hands-on, thought-provoking and collaborative in nature. The labs were based on problem-solving and data-driven activities and designed for use with the participants' students. They also were created using easily accessible and affordable materials.

One of the most important aspects of a successful classroom experience is to have *fun*. Throughout the workshop lessons, this is a goal. Check out this video of Workshop 1 and Workshop 2 (https://www.youtube.com/watch?v=hhZ8UVPbVdU&t=29s) and see an active classroom full of engaged students (teachers in this case) learning and having fun.

Workshop 2: Snow and Ice Science

The focus of Workshop 2 centered on snow and ice science, while reinforcing the inquiry/constructivist skills introduced in Workshop 1. This workshop included access to cutting-edge research, lab experiences, scientist presenters, and field experiences designed to increase educators' background knowledge. Classroom activities and labs were created to engage learners and to transfer current ice science and climate change concepts to classroom curriculum. The following will be an annotated list of activities available to you on the Live Agenda.

The first day of Workshop 2 introduced the polar regions through an activity called "To the Ends of the Earth." Living in Greenland, our audience was obviously well-versed in the Arctic, but this activity compared the Arctic to the Antarctic and encouraged discussion of their own experiences living in the extremes of their polar region. It also allowed discussion about the special needs of researchers working in the field in both Greenland and Antarctica.

To introduce the principles of ice core science, workshop attendees built a "Life Core," where art meets science. The Life Core, built using a variety of materials, told the events in each participants' life that led them to becoming

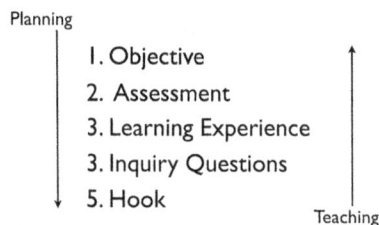

Planning

1. Objective
2. Assessment
3. Learning Experience
3. Inquiry Questions
5. Hook

Teaching

Figure 2.5 Backward design lesson planner

Figure 2.6 Life core with repeating layers

Figure 2.7 Sharing a life core

Figure 2.8 Life core with thin and thick layers

a teacher. Like a real ice core, their Life Core had a top layer that was the most recent event and went back through time to the oldest layer at the bottom of the core. Like ice from different years, some layers are thicker or thinner than others and the layers can be repeated, and some layers might even be missing.

Introducing "Snowflake Science" also began with an art activity and a challenge. Students explored cutting snowflakes and then were challenged to cut a hexagonal (six-sided) dendritic (branched) snowflake. The paper snowflakes were then used to explore the falling rates of snow and compared their paper models to real snowflakes' falling rates.

"Snowpit Science" incorporated several of the skills and tools developed throughout this workshop to demonstrate how teachers can provide an opportunity for their students to develop their own inquiry questions and investigations.

For Workshop 2, a new resource being developed by IDP was introduced. "Virtual Field Labs" (VFLs) grew out of a need during COVID-19 to have excellent science resources that could be used in the classroom or independently by students on their home computers/tablets, or in a hybrid learning situation. VFL's are facilitated investigations that engage students in generating, analyzing and interpreting data that parallels current IDP research initiatives. VFLs include the following.

1. Hands-on virtual tools for teaching climate change concepts.
2. Student-generated data activities led by climate change scientists.
3. Generating, graphing and analyzing data on the way to making claims supported by evidence related to the investigation and connections to the future of climate change.

Two examples of VFLs were used in the workshop and are linked in the Live Agenda. More VFLs have since been created based on additional climate change topics and can be found at (https://icedrill-education.org).

"Direct Measurements: Tree Rings and Ice Cores" leads students to investigate the meaning of data that comes from "direct measurements." The directions for making the ice cores used in this lab can be found at https://icedrill-education.org/portfolio/decoding-ice-cores-proxy-and-direct-measurement-2-labs/. Another 'direct measurement' lab is "Decoding Ice Cores: Atmospheric Analysis of CO_2 Record" in which students using a TDS (total dissolved solids) meter sample model "melt water from different levels from an ice core" and graph their findings of "CO_2." They then compare their graphs to the real data in the 800,000-year record of CO_2 in the atmosphere.

On the third day of Workshop 2, we investigated models. By using "Ice Balloons," students investigated a physical model of glaciers and moulins. Through "Slip Sliding Away," students explored glacier dynamics and developed investigable questions. Moving from direct measures to proxy measures, students completed the "Decoding Ice Cores: Developing Proxy Measures" lab. Isotopes in the ice cores were used to indicate what the temperature was back in time. The "Decoding Ice Cores: Isotopes" lab takes students through the use of a proxy measure. The "Nature vs. Science" hypothesis card game placed pairs of students in the roles of "nature" and "scientist(s)" in playing this game. The game illustrates for students how scientists test hypotheses to eventually develop a theory.

"Land Ice/Sea Ice/Grounded Ice" helps students answer the question of "How will melting ice affect sea level?". The participants explored a model of a marine-terminating glacier to understand how upwelling of freshwater (meltwater) in front of the glacier influences the ecosystem of a fjord. Using easily attained materials, students engineered a drill rig in an activity titled "Engineering Challenge: Designing a Drill Rig." Besides learning about engineering, students also gained an appreciation for the challenges engineers face when having to create a drill rig that can be moved across the frozen landscapes of Greenland and Antarctica.

"Thermohaline Circulation" demonstrations provided experience for students to understand how different densities from temperatures (thermo-) and salt concentrations (-haline) affect global ocean circulations. "Albedo and Feedback Loops" provided an engaging activity for students to investigate temperature and albedo. This is also an activity that lends itself well to setting up longitudinal studies with your students.

COVID-19 Impacts

SNOWI was originally planned as a two-part series for in-service and pre-service teachers, with one workshop in the fall of 2020 and the second in the spring of 2021. Due to COVID-19, the two-workshop program was postponed an entire year. The materials and agendas were reorganized to be presented in a virtual format. This was necessary because the original grant funding included tasking the pre-service teachers with using lessons from the workshops as part of their student teaching requirements in the fall semester of 2021. Their Workshop 1 was primarily focused on inquiry teaching and constructivist learning pedagogy and skills demonstrated with hands-on science experiences. The COVID-19 hybrid model with the pre-service teachers took place in September 2021.

Many of the in-service teachers working with the university in Nuuk come from distant settlements and towns scattered along Greenland's immense coastline. The original plan was to have a workshop in the fall and one in the spring, with time between the two workshops when teachers could use the materials and skills from Workshop 1 with their students. COVID-19 required us to pivot to a different model in which the workshops for the in-service teachers were presented back-to-back. This allowed the traveling teachers to only have to travel once.

Before beginning Workshop 2, the plan was for the in-service teachers to arrive on the weekend between the workshops to experience an abbreviated version of the content from the first workshop, then both groups of teachers would come together and engage in more in-depth snow and ice content during the week of Workshop 2. Unfortunately, COVID-19 did not allow things to smoothly follow even our careful hybrid plan. There were 13 pre-service teachers enrolled in Workshop 1. All were second-year education students at the University of Greenland. Actual attendance was difficult to quantify because it changed rapidly, often several times in a day.

The presenters led activities and discussions on Zoom, and the healthy students met in person in the classroom with their professor. COVID-19–quarantined and ill students joined synchronously on Zoom, or in some cases accessed the recorded meetings later.

Students met with their professor in the morning and discussed readings and the prior day's hands-on activities. He was in direct contact with all students each day and helped the ill ones attempt to keep up. The students able to attend class worked with him to set up the lab materials for the afternoon meeting. By engaging the participants in setting up and cleaning up the hands-on activities, they gained knowledge on how to manage materials when they are in the role of teacher.

After lunch, students attended the workshop led by the Zoom presenters who modeled inquiry teaching while leading them in classroom lab activities they can use with their students. Each new learning activity was carefully scaffolded to build upon earlier ones, so that by the end of the workshop, the participants internalized a clear understanding of what inquiry teaching is and how to create strong constructivist experiences for their students.

Workshop 2 for both pre-service and in-service teachers took place in March 2022. The abbreviated version of Workshop 1 for the in-service teachers was planned for Friday and Saturday before Workshop 2 which would begin on Monday for all participants. The pre-service teachers who had been ill during the fall Workshop 1 asked to attend again with the in-service teachers to catch up on what they had missed while sick. In the end, *all* of the pre-service teachers attended a second time because only three students had been able to attend every day in the fall. Those three wanted to attend again for several reasons: 1) they had really enjoyed the experiences; 2) they felt hearing it again would help them process all that they had learned; and 3) with all of the comings and goings of their cohort, it had been a bit disjointed for everyone.

Challenges

1. **COVID-19** affected every aspect of these workshops and caused two years of delays and complications in planning and carrying out the workshops. Flexibility was required, as was the ability to pivot quickly to meet the needs of the participants.
2. **Language** was a huge challenge because almost all students spoke at least three languages. They either spoke Greenlandic or Danish as their first language and the other was their second language. English was their third language, and in some cases, they did not speak English at all. All Zoom presenters spoke only English. The professor spoke Danish and English well, and understood most in Greenlandic.

   a. Readings were translated into Danish beforehand and a few additional were translated during the week of the workshop. The professor decided Danish was the best language for this workshop. He translated materials and gave verbal directions in Danish. Students helped translate into Greenlandic for the students who did not speak English or Danish well.
   b. We used Google Translate for adding Danish to our slide presentations, and then the professor checked the translations for accuracy.
   c. Directions were built into Keynote and PowerPoint slides with as many visual graphic representations as possible. The professor translated verbal directions as needed.
   d. Presentations used visual props to demonstrate ideas.
   e. Presenters worked hard to recognize English words and phrases that would not translate easily to Danish, and to be aware of using slow, deliberate speech when presenting.
   f. Interesting observation: by the third day, oral translations were not needed as often – students' "ears" became accustomed to presenters' speech and accents.
   g. VFLs were accessed through YouTube and incorporated Danish subtitles.

3. Materials had to be shipped from the United States months ahead of the workshops. When the workshops were canceled because of COVID-19, the materials had to be stored in their shipping boxes for about two years. Careful and thorough inventories had been made when the materials were packed, but issues arose, including some corroded batteries. The transformation from an in-person workshop to a hybrid one included some changes in agendas. The addition of new labs required adjustments in materials available in Greenland.

Evaluations

Even with all of the COVID-19 impacts on attendance, the evaluations came back with very positive comments.

When the participants were asked if they felt as if they had acquired new teaching competencies, 100% responded that they had. In their words, the new things they learned included the following.

- Fun teaching.
- Different ways to teach.
- Engineering (it is *sooo* fun).
- I have learned that IBSE (inquiry-based science education) can be a very good tool to teach students. Students can better understand subject matter and practical activities.
- Constructivism.
- I have learned how to include and listen to students, so they can know or feel that I am there to help them.
- How I can develop students' interest/curiosity.
- IBSE – and backward planning – made me think of new ways to teach science.

When asked how well the virtual hybrid model of presentation worked, they responded with the following.

Things That Worked

- Collaboration.
- Flexibility in teaching. both on screen and in the classroom.
- That you could be at home and still learn.
- Though I was home, it was good to join in online. I envy the others, but am happy to have been a part of the week.

Things That Could Have Been Better

- If they (Louise, Bill and Jim) were here.
- It could have been better if the Americans were here.

- It would have been better if they were here.
- That I knew it was the same Zoom link for all days.

  (Author's note: it was the same Zoom link and was listed on the Live Agenda, but communication with the ill students was sometimes difficult, so it is easy to understand some confusion.)

- First time on Zoom.

When asked, "What have you learned about teaching in multiple languages in a classroom?" They responded as follows.

- Translation is good – only lost a few words.
- Three languages are a bit difficult, but it worked.
- Words disappear – my tongue doesn't work.
- Sometimes it is hard – my English is not very good.
- Good – if you don't understand the explanation, it can be translated by someone in the room.
- I have gained many ideas for teaching.

Conclusion

The lessons from Greenland showed that teaching snow and ice science to future teachers in Greenland proved to be a very rewarding experience for educators and participants in the workshops. The pre-service teachers who participated in the workshops began their third year teaching practice period in the first week of January 2023. All of them decided to include one or two activities from the workshops in their teaching practice. Based on talks with the participating pre-service teachers, it is difficult to underestimate the value the workshops have had on their future teaching practice.

Creation and facilitation of excellent professional development (PD) can be a challenge in the best of times when it comes to professional development for teachers. COVID-19 caused years of the necessity for educators everywhere to be flexible and creative in finding solutions to providing high-quality learning experiences for students. The lessons learned from leading an international teacher PD workshop at the height of COVID-19 may provide ideas and usable resources for educators in a wide variety of teaching situations.

In terms of recommendations from this experience, there are the following key take-home messages.

1. It does not matter if we are involved in polar education, virtual learning or classroom situations – good teaching involves learning experiences that engage students in active learning.
2. The hybrid workshop model presented in this chapter had one presenter in the room with the participants, while other presenters were on the

"Zoom wall." The presenter in the room was essential to the success of this workshop because he was able to "read the room" and gauge whether the participants were understanding the presentations and directions. He was able to call a "timeout" at any point for a chance to clarify or for participants to take a break. Because of the many hands-on activities and the fact that there were multiple first languages in the room, we cannot stress enough how important this was.

3. In general, polar science content is often presented through a lecture format, but this workshop illustrates how rich the learning experiences are when students' prior knowledge is activated through an inquiry activity before presenting new concepts (ABC: activity before content).

4. And do not forget to have *fun*!

References

Anderson, L.W., Krathwohl, D.R. and Bloom, B.S. (2001) *A Taxonomy for Learning, Teaching, and Assessing: A Revision of Bloom's Taxonomy of Educational Objectives*. Complete edition. Longman, New York.

Bodner, G. (1986) *Eternal Verities [Brochure]*. Purdue University Chemistry Department, Lafayette, IN.

Bowen, R.S. (2017) 'Understanding by design', Vanderbilt University Center for Teaching, https://cft.vanderbilt.edu/understanding-by-design/ (accessed 12 December 2022).

Chu, S.K.W., Reynolds, R.B., Tavares, N.J., Notari, M. and Lee, C.W.Y. (2017) *21st Century Skills Development Through Inquiry-Based Learning from Theory to Practice*. Springer, Singapore.

Ice Drilling Program. (2023a) 'IDP long range drilling technology plan 2023', https://icedrill.org/long-range-drilling-technology-plan

Ice Drilling Program. (2023b) 'IDP long range science plan 2023', https://icedrill.org/long-range-science-plan

Jones, G.M. and Brader-Araje, L. (2002) 'The impact of constructivism on education: Language, discourse, and meaning', *American Communication Journal*, 5(3) (Spring).

Pedaste, M., Mäeots, M., Siiman, L., de Jong, T., van Riesen, S., Kamp, E., Manoli, C., Zacharia, Z. and Tsourlidaki, E. (2015) 'Phases of inquiry-based learning: Definitions and the inquiry cycle', *Educational Research Review*, 14.

Pierce, W. (2001) 'Inquiry made easy', *Science and Children*, 39–41.

Seag, M., Badhe, R. and Choudhry, I. (2020) 'Intersectionality and international polar research', *Polar Record*, 56, e14. https://doi.org/10.1017/S0032247419000585.

Walbert, D. (2014) 'The learning cycle', https://inquiryteaching.weebly.com/uploads/1/0/3/3/10332398/learning_cycle_2014_3_phase.pdf

Wiggins, G. and McTighe, J. (2005) *Understanding by Design*. 2nd edition. Association for Supervision and Curriculum Development, Alexandria, VA.

Willard, T. (ed). (2015) *The NSTA Quick Reference Guide to the NGSS K-12*. NSTA Press, Arlington, VA.

3 Ecology as the Foundation for Understanding and Integrating Basic Sciences

Anne Farley Schoeffler

The study of ecology is based on basic sciences, including – but not limited to – chemistry, biology, and meteorology. Thus, it should be taught in conjunction with basic science rather than in isolation. In fact, student engagement and understanding of science concepts benefit when their focal point is the natural world; they are better able to comprehend basic sciences when these relate to students' experience of the world outside their doors. Specifically, science classes in middle school (12–13-year-old students) are designed to relate curricular standards throughout the year to the relationship of those phenomena with the environment.

Early in the academic year, students review basic skills (data collection, analysis, graphing) using the data they collect themselves outdoors (e.g., insects, flowers, etc.) before launching into more challenging concepts. Their chemistry lessons (atomic structure, matter, reactions) are related to environmental processes. Their energy lessons, likewise, focus on conversions beginning with sunlight and relating to atmospheric and meteorological consequences, especially as those contribute to weather and climate. With a foundation in global patterns, students explore biomes and relate them to geographical and meteorological criteria and existential threats to biomes.

Introduction

Freshwater is an essential resource, often undervalued in the recent past and increasingly strained in the present. It is cyclically reintroduced to the landscape primarily via precipitation and moves from higher to lower elevation as a result of gravity. Consequently, it runs off impermeable or saturated surfaces towards lowest elevation areas within the confines of the highest elevation boundaries; this comprises a watershed. The common outlet which drains a watershed may be a small stream, river, lake, bay, or other body of water. Moreover, the local watersheds of tributary streams are subsumed into the regional watersheds of larger bodies of water; thus, smaller watersheds combine to form larger watersheds.

DOI: 10.4324/9781003486961-3

Human behavior often negatively affects the quality of water, which has a cumulative effect on larger watersheds. In particular, pollutants that are directly introduced to surface waters (point source pollutants), those that are introduced via diffuse sources (nonpoint source pollutants), and increased sedimentation as a result of deforestation or construction may enter water bodies in smaller and/or larger watersheds. Those that begin in smaller tributaries do not remain confined to those, however, but flow into the larger water bodies as a natural course. The effect of pollutant loading of this kind is to increase the level of contamination in larger water bodies, resulting in seriously damaged riparian ecosystems and dead zones in marine coastal ecosystems. Furthermore, one consequence of global warming is the increased water flow in Arctic and alpine ecosystems as a result of increased glacial melt. Changes to these watersheds have ecological repercussions, as well as potential effects for human communities in downstream regions.

It is thus incumbent upon educators to build students' conceptions of watersheds as systems, as well as the sustainable exploitation of those systems for human consumption, agriculture, and industry. The following set of lessons is designed for middle-level students, ages 11–14. They follow the 5E's instructional model – Engage, Explore, Explain, Elaborate, Evaluate (**1**) – and are designed to be completed in 5–10 40-minute class periods, depending on which activities are selected. The 5 E's cycle is intended to allow students to investigate a topic and build understanding through their own observations and actions with scaffolding supplied by the teacher at significant points in the process. As they build core concepts and vocabulary, they can then apply these to related phenomena for deeper understanding (**1**).

Engage

An Engage activity within the 5E's instructional model

> mentally engages students with an activity or question. It captures their interest, provides an opportunity for them to express what they know about the concept or skill being developed and helps them to make connections between what they know and the new ideas (**1**).

The activities relayed here introduce students to the structure and mechanics of surface water flow, erosion, and the basic vocabulary of watershed components. The investigation (which may also be conducted as a demonstration if there are limited supplies) can be completed in one 40-minute class period, or may be extended in various ways as indicated.

Students use a stream table to model stream mechanics. Stream tables are commercially available but can also be created with the use of a large tray (at least 1 meter in length), blocks to raise one end, and a means to manage overflow (see Figure 3.1). These most often hold sand and have water added by means of a spigot or a pitcher. Ideally, the bottom one-third or so of the

Figure 3.1 Stream table
Photo: Schoeffler

space is left empty as it will form a lake or ocean. Another alternative is to use a sloped, sandy, outdoor area.

As an Engage activity, students carve an initial stream into a sandy slope and then pour water into the headwaters or source of that stream. The water will carve the initial channel deeper, possibly carve additional channels, create a floodplain, and possibly also create a delta. Students draw a two-dimensional map on paper of their water system and then identify and label basic vocabulary (river source, floodplain, meander, delta, tributary); the definitions of these terms may be provided or sought in a textbook or online. (It is possible that students also indicate contour lines if they have a previous understanding of topography.) This exercise directs students to systems thinking as they relate topography (the slope of the land) and gravity to water flow and erosion. Water flows downhill as a result of gravity, and river flow is not directed by compass directions; conceptually, many students struggle with the idea that north does not mean "up".

A further investigation of the students' streams sees them developing model towns along the banks, together with engineering solutions for maintaining

problem.

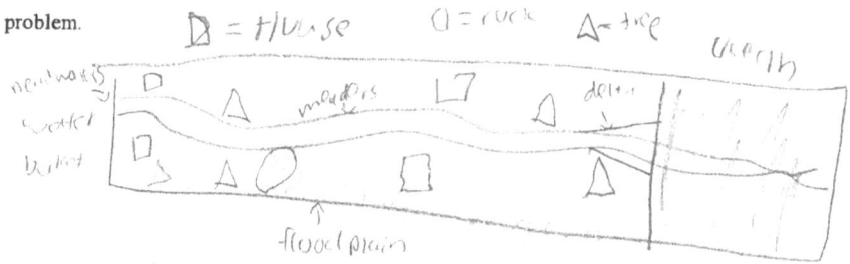

Figure 3.2 Student stream table map
Source: Schoeffler

the structures despite flood conditions. Students are supplied with a number of buildings (which can be made of Lego bricks or be Monopoly houses, blocks, or structures of their own design), which they add to the stream model. They are also given a limited budget in imaginary money to apply towards protecting the homes using rocks, plastic plants (for example, aquarium plants), or other objects of this type. Students then "plant" the trees or construct barriers that should mitigate flooding and preserve the structures. They test with more water flowing downstream and then evaluate the results: Did they save their buildings? What solutions were the most cost effective?

These Engage activities, probably comprising one or two 40-minute class periods, help the students build and model basic vocabulary; relate topography to runoff, stream flow, and erosion; and give context to the relationships between water systems and human habitation. They will help students visualize regional watershed systems in the context of local experiences.

Explore

In the Explore phase,

> students carry out hands-on activities in which they . . . explore the concept or skill. They grapple with the problem or phenomenon and describe it in their own words. This phase allows students to acquire a common set of experiences that they can use to help each other make sense of the new concept or skill (**1**).

In this case, students will explore the movement of water on and around familiar structures and identify the direction in which rainwater flows away from buildings and over the landscape. They will begin to see the relationship between their own experience of runoff and the larger watershed system.

An in-class lesson introduces students to a historical water system, such as that of the 15th century Inca at Machu Picchu in modern-day Peru. That mountaintop city was constructed with ingenious channels and downspouts

Figure 3.3 Machu Picchu
Photo: Schoeffler

designed to deposit rainwater in gardening terraces and also to direct the surplus downhill and away from residences. For an explanation, see "Machu Picchu's Remarkable Water Supply and Drainage Systems" (2).

After that introduction to an ancient drainage system, the class begins an outdoor walking activity in which they trace the drainage from their own school buildings and campus. Using a plot of the property, students mark in red all the apparatus used for the purpose of draining water, such as downspouts, drains, ditches, retaining ponds, etc.; see the sample in Figure 3.4. A plot of this type may be obtained from the school office or development department, or it may be obtained from an aerial image, for example as found on Google Maps (which can be traced, so that it shows outlines only and not vegetation, etc.). The class will discover that water is typically directed through horizontal gutters to downspouts which then direct the water away from the building; drains may be visible in the parking lot where downspouts tie in. Water may flow off campus through storm sewers, ditches, culverts, and/or into a retention pond that subsequently directs it through additional ditches, culverts, or storm sewers. Most students have not considered how water moves, especially during significant rain events.

After exploring the rainwater systems on campus, students are directed to investigate the system(s) associated with their homes. They use digital images

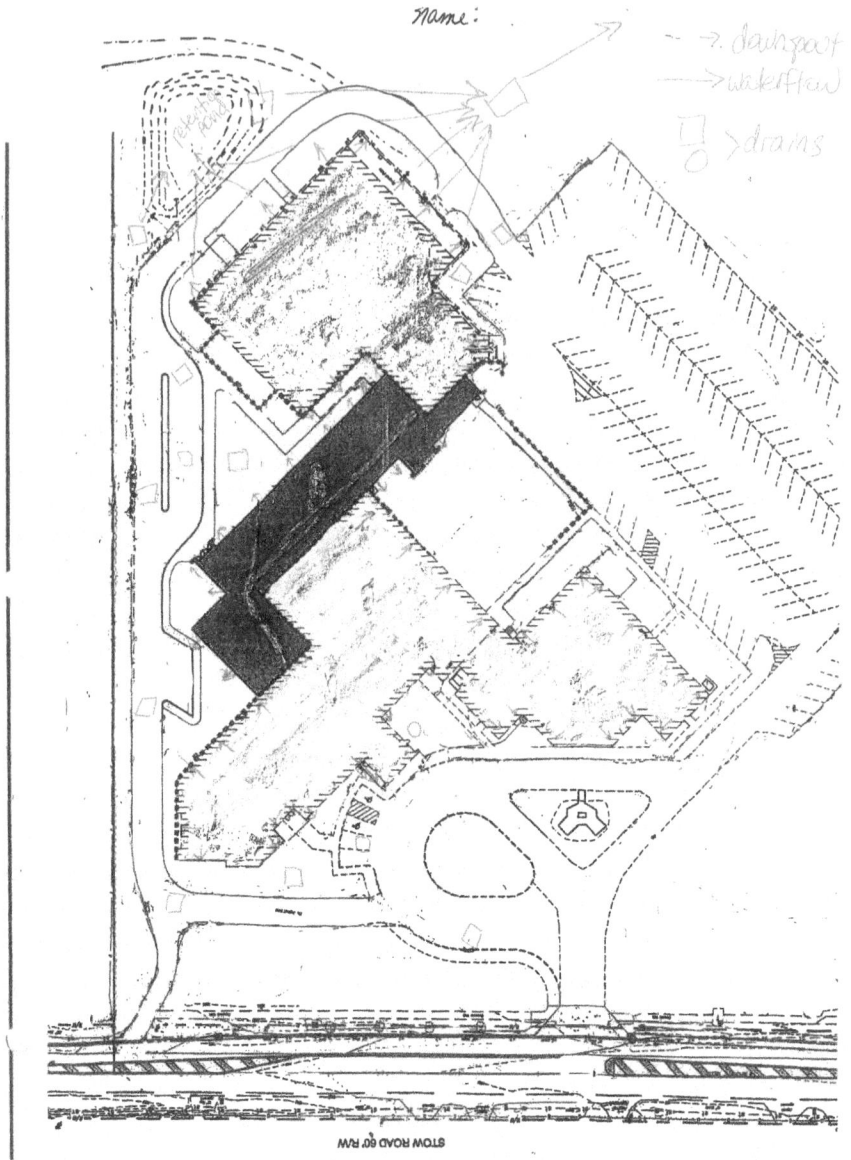

Figure 3.4 School plot with drainage student notation for drainage
Source: Schoeffler

to show where water is collected from the roof (gutters and downspouts), where it runs off across the ground, and where it leaves the property (drains, ditches, streams, retention basins). Students are directed to begin with a 70–100-word summary of the system as a whole, include 5–8 photos, and provide a caption for each photo of only 15–25 words. These guidelines were designed with reference to National Geographic's *Story-Telling for Impact in Your Classroom: Photography* course for educators (3). An alternative home-based assignment would be for students to map the home and waterflow patterns similarly to the map they annotated for their school. In either case, students relate their experience of the stream flow model (Engage activity described previously) to the flow of water in their own local circumstances.

In another outdoor walking activity, the class visits a surface water system, in this case a small lake, in order to map locations where water enters and leaves the lake. Using a map or aerial image of the water body, students annotate that document with notation showing where water enters and leaves, and the direction of flow. An alternative activity would be to help students annotate the map or image without actually visiting the location. If the map includes contour lines, students can use those to identify the downward flow of runoff with the topography of the surrounding landscape. They are able to visualize the relationship between elevation and runoff, necessary information for understanding the nature of a watershed. This opportunity can double as an ecology exploration in which students also investigate trees, wildflowers, insects, and/or amphibians.

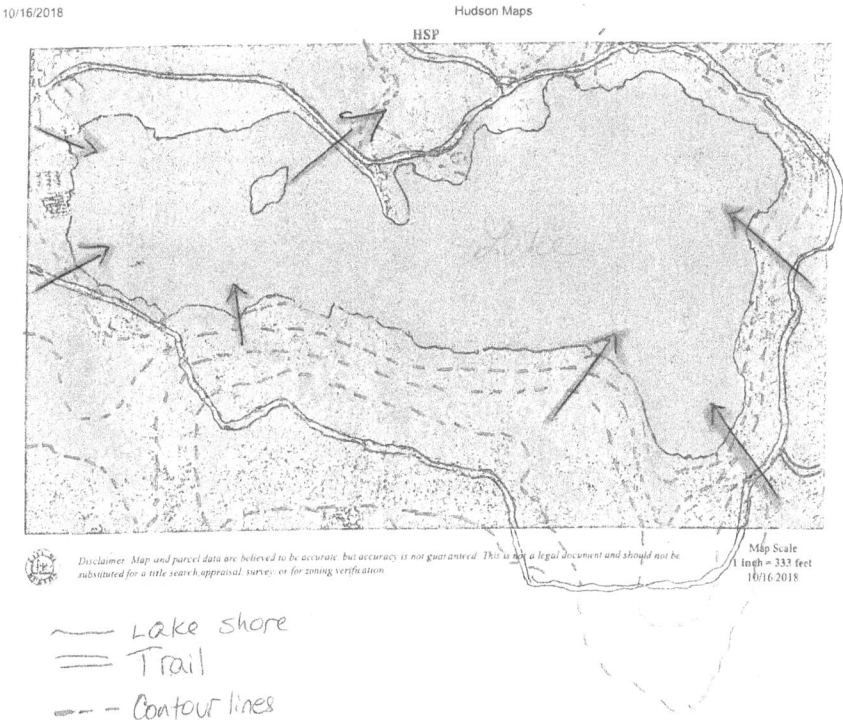

Figure 3.5 Hudson Springs Park map with contour lines and student notation for inputs and output

Source: Schoeffler

These Explore activities could be completed in 1–2 classes or days, although the second walk is likely to take longer than a class period unless the water body is very nearby and not very large.

Explain

The Explain stage of the 5E's model is the point at which "the teacher provide[s] concepts and terms" to assist students in "develop[ing] explanations for the phenomena they have experienced" (1). In this particular educational cycle, the teacher defines the terms "watershed", "surface water", "ground water", and "aquifer", and explains the mechanics of a watershed. This lesson would be completed in a single class period or less.

Watersheds are divided and subdivided into successively smaller hydrologic units (HUs) (4). For example, the U.S. Geological Survey classifies the landscape into levels 1–6: regions, subregions, basins, subbasins, watersheds, and subwatersheds. The largest designation, the region, is a major land area; the Lower 48 United States comprise 21 regions (5).

The watershed level is typically 40,000 to 250,000 acres, and the subwatershed level is typically 10,000 to 40,000 acres with some as small as 3,000 acres. Approximately 18,000 watersheds and over 98,000 subwatersheds are mapped to the 5th and 6th level (5).

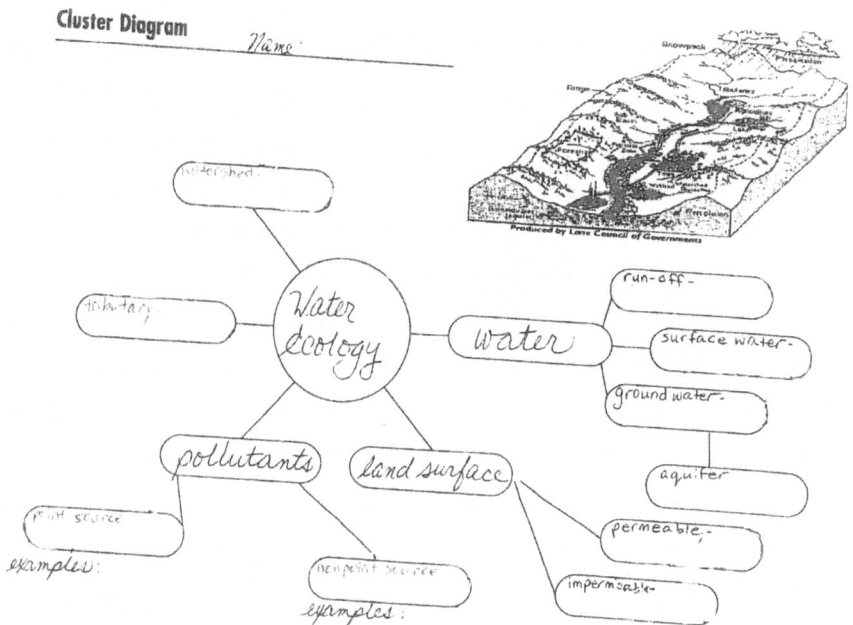

Figure 3.6 Water graphic organizer

Source: Schoeffler

Thus, a small, local watershed is always a tributary to a larger one.

Students use local and regional maps to relate their previous experiences to a subwatershed and then a watershed. For example, using a map such as the one shown in Figure 3.7, students can identify cities and streets that they recognize. They can also relate their walking experience to the watershed in which that occurred. Subsequently, students use a watershed map to

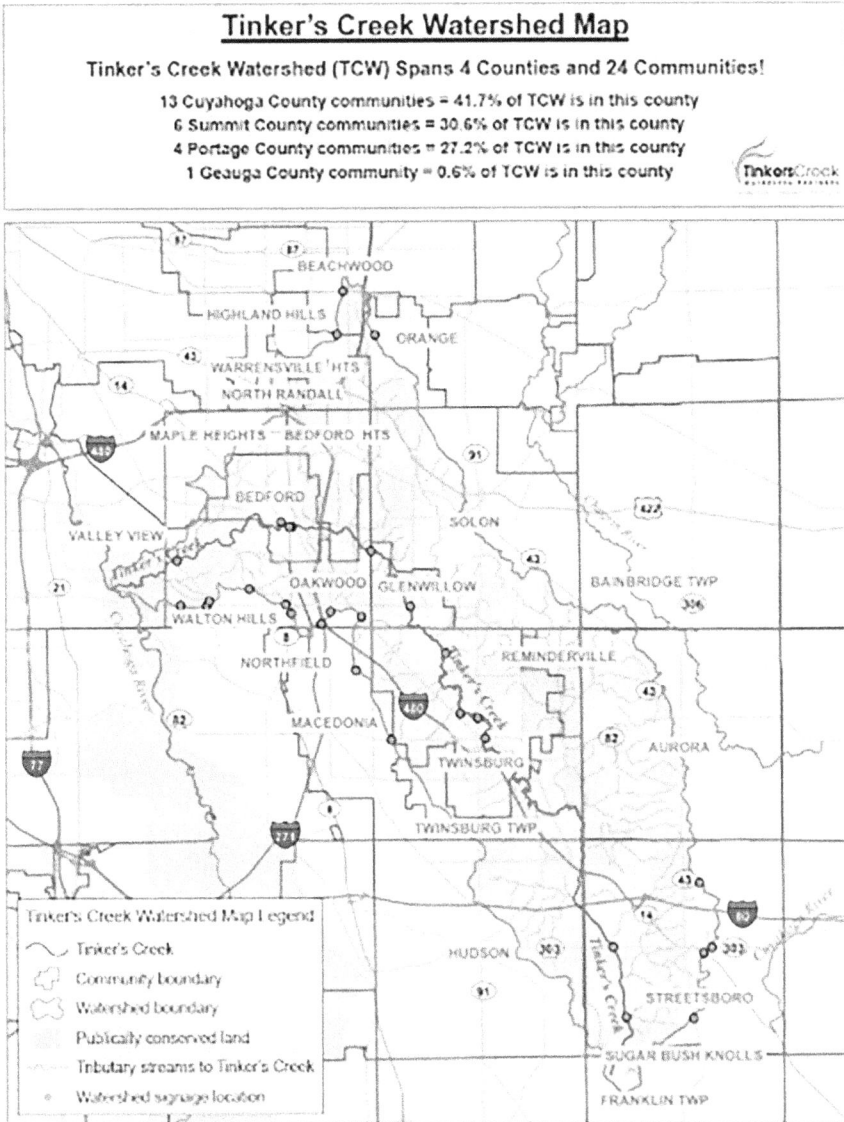

Figure 3.7 Tinkers Creek Watershed Map

Source: Tinkers Creek Watershed Partners (6)

localize the subwatershed with which they are familiar into the larger region. In this particular case, the smaller Tinkers Creek subwatershed contributes to the larger Cuyahoga River Watershed, which is part of the Lake Erie Basin, which in turn is a part of the Great Lakes Basin. Ultimately, all of these water bodies drain through the St. Lawrence River to the Atlantic Ocean.

Elaborate

In this phase of the 5 E's model, students have opportunities to apply what they have learned and further develop their conceptual understanding or skills. Once they have an understanding of the movement of water through a system, especially a watershed, they can investigate pollutant loading, whether from point or nonpoint sources. Sedimentation in a water body is a relatively simple process to identify: turbidity (the cloudiness of the water) is tested using a Secchi disk (Figure 3.8). Invented by Angelo Secchi in the nineteenth century, the disk is simply lowered into a water body until it is no longer visible; the limit of visibility is then measured in centimeters and converted into Nephelometric Turbidity Units (NTUs) and thence to a Q-Value akin to a water quality score. The graph shown here is modified for a table-top scale from Mitchell and Stapp's *Field Guide for Water Quality Monitoring* (7).

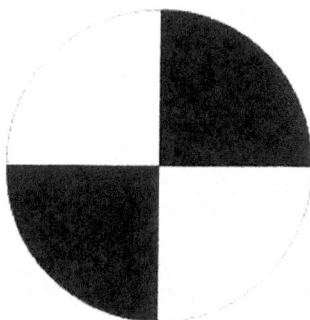

Figure 3.8 Secchi disk

Turbidity testing makes an excellent field experience; however, it may prove impractical to take students to multiple locations in a short period of time. An alternative is to do table-top Secchi disk tests, either with locally sourced samples or with artificial samples. A set of identical containers (such as tennis ball tubes wrapped in white paper) treated with varying drops of chocolate milk (5–35 drops) will create a diversity of turbidity samples (8). The scaled table and graph help students convert their measurements to NTUs and assess the "water quality" for the samples; see the table here. Sample water body names include: Chattering Mountain Stream, Holiday Lake, Upper Micro River, Lower Micro River, Upper Mighty River, Lower Mighty River, and Grey Bay. The samples should be intentionally created with the specific objective of increasing turbidity throughout the watershed; specifically, the Chattering Mountain Stream will be clearer than the Micro River, and the Lower Mighty River will be the most turbid, receiving turbidity from all tributaries as it would do in a natural watershed. The calm water bodies (Idyll Pond, Holiday Lake, and the Grey Bay) allow sediment to settle and are thus less turbid than their tributaries. This activity can be completed in 1–1½ class periods.

Observations:

Water Body	Depth (cm)	NTU	Water Quality Grade
Chattering Mountain Stream			
Idyll Pond			
Upper Micro River			
Lower Micro River			
Holiday Lake			
Upper Mighty River			
Lower Mighty River			
Agua Bay			

Figure 3.9 "Using a Secchi Disk" lab data table

Source: Schoeffler

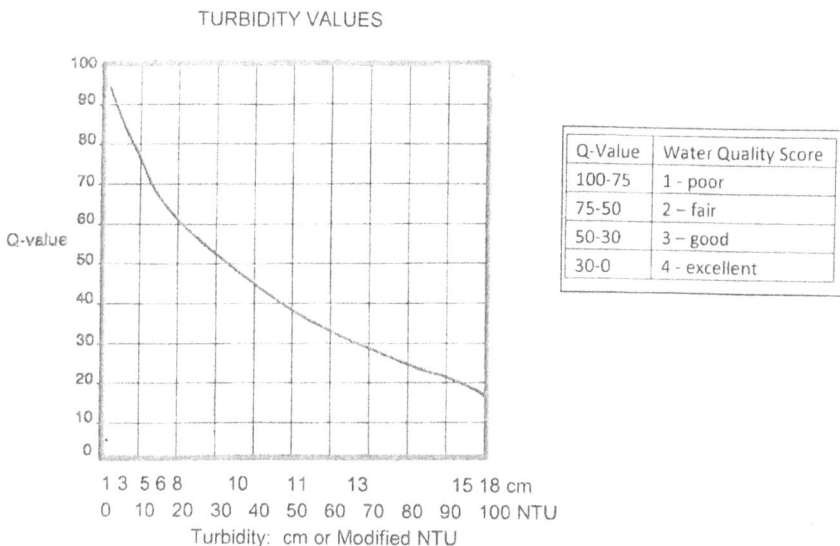

TURBIDITY VALUES

Q-Value	Water Quality Score
100-75	1 - poor
75-50	2 – fair
50-30	3 – good
30-0	4 - excellent

Figure 3.10 Graph scaled to mini table-top Secchi disk activity

Source: Schoeffler

Observations

Moreover, students then use their findings to consider a sequence of tributaries, proposing a watershed structure whereby a map can be drawn showing the cleaner tributaries higher in the watershed, the more turbid ones lower, and a somewhat cleaner outlet at the end to an ocean (as it can dilute the accumulated pollutants). This final map-drawing stage is conceptually challenging for students of this age. A further demand might be to require that they also include contour lines representing topography that shows higher elevation higher in the watershed and lower elevation at surface water shorelines.

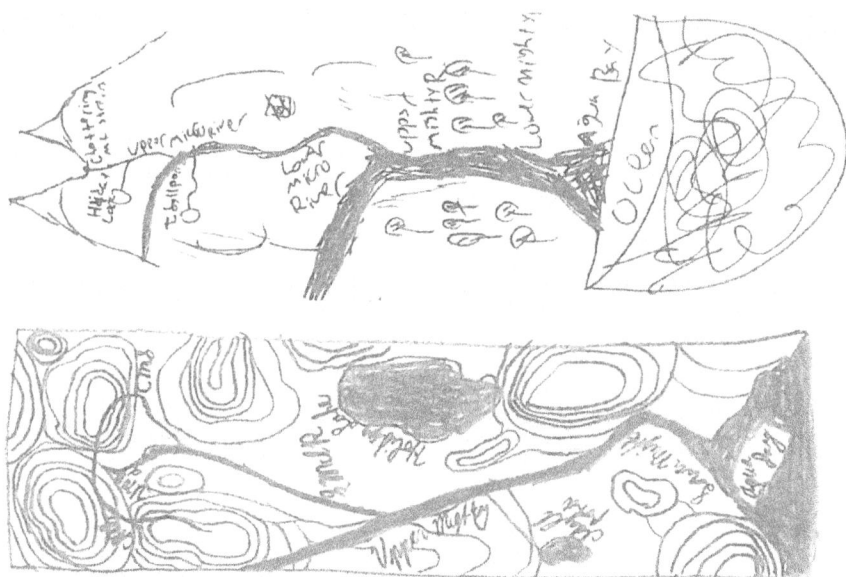

Figure 3.11 and Figure 3.12 Student maps for the watershed they conceive after testing the artificial samples (Schoeffler)

Evaluate

The final phase "provides an opportunity for students to review and reflect on their . . . learning, . . . understandings, and skills," (1) and to provide evidence of their learning. In this instance, the objective is to relate their understanding of a local, familiar system to a global context. Students or student groups research a problem associated with a tundra watershed and may also propose a potential solution to the problem. Students select an Arctic locale or a "third-pole" alpine tundra issue. Some potential areas of concern include: downstream flooding, pollutant loading, restricting flow (damming), damn breaks, invasive species, etc. The project selected requires that students understand the structure of a watershed and the impact that tributaries have on a larger water body and/or a downstream landscape. For example, glacier runoff that initiates a river is subject to increase as the climate warms, increasing turbidity downstream. Swelling rivers can also cause glacial lake outburst floods (GLOFs) that can have devastating downstream effects. Introduced species can spread through a watershed by means of streamflow. (Some of these topics may be the focus of additional lessons prior or in addition to this watershed unit. For example, there may have been earlier lessons relating to invasive species ecology or glacial outflow.)

The form this project takes should be subject to the students' decisions; some ideas include: a slideshow, mural, story book for younger children, three-dimensional model, research paper, podcast, interview-style "talk show," or a webpage. Students' engagement is likely to increase commensurate with their ability to choose aspects of the project (**9**), including the watershed of interest, the topic, and the product they create. These activities are liable to take several days if that time is available or can be scaled back to suit the time constraints of individual classrooms. Students conclude by presenting their findings to their peers and/or other stakeholders. Ideally, in creating and revising their work, students peer edit one another's projects prior to revisions, assess one another's presentations, and self-assess their own productivity.

A summative exam can also be used at this juncture. Evaluate-activities (projects, exams, etc.) should be designed so that students are able to explain what a watershed is, how runoff is directed within a watershed, the structure of watershed system composed of its tributaries, and thus how pollutant loading occurs in increasingly larger hydrologic units. Students relate their experience of a local system to an Arctic or alpine system, as those are highly affected by the Earth's warming climate with repercussions on local and global ecological and human systems.

References

(**1**) Next Generation Science, *5 E's of Science Instruction*, viewed 27 November 2022. http://nextgenerationscience.weebly.com/5-es-of-science-instruction.html

(**2**) Omrania, *Machu Picchu's Remarkable Water Supply and Drainage Systems*, viewed 27 November 2022. https://omrania.com/inspiration/machu-picchus-remarkable-water-supply-and-drainage-systems/#:~:text=The%20city's%20walls%20of%20heavy,drain%20runoff%20into%20the%20ground

(**3**) National Geographic Online Courses for Educators, *Storytelling for Impact in Your Classroom: Photography*, National Geographic, completed 17 June 2021.

(**4**) United States Geological Survey 2022, *Hydrologic Unit Maps*, United States Geological Survey, viewed 27 November 2022. https://water.usgs.gov/GIS/huc.html

(**5**) South Dakota State University Extension 2022, *What Is a Hydrologic Unit Code?*, South Dakota State University, viewed 27 November 2022. https://extension.sdstate.edu/what-hydrologic-unit-code-huc

(**6**) Tinkers Creek Watershed Partners 2022, *Tinkers Creek Watershed Map*, viewed 27 November 2022. https://tinkerscreek.org/

(7) Mitchell, MK and WB Stapp 2008, *Field Manual for Water Quality Monitoring*, Kendall/Hunt Publishing Company, Dubuque, IA.

(8) Schoeffler, AF 2017, 'Testing the Waters', *Science Scope*, NSTA Press, vol. 40, no. 6, pp. 80–83.

(9) Larmer, J, J Mergendoller and S Boss 2015, *Setting the Standard for Project Based Learning*, ASCD Alexandria, VA and Buck Institute for Education, Novato, CA.

4 Using Place-Based Education to Teach About Climate Change and Connect to the Arctic

Regina Brinker

Introduction

Climate change is happening most rapidly in the Arctic. Temperatures are warming up to four times faster in the Arctic as compared to other areas (Jacobs et al., 2021). Changes in polar regions influence environmental, health, social, economic, and political systems globally. Because of these impacts, it is important for us to understand how we are connected to the Arctic, regardless of where on Earth we live.

Changes in the Arctic affect global economic, political, social, and environmental systems. Changes may be seen as positive or negative, depending on your point of view. For example, a decline in arctic sea ice opens shipping routes for travel, tourism, and resource exploration. More travel, however, increases pollution. Loss of sea ice reduces wildlife habitat, increases coastal erosion and decreases polar albedo, or reflectivity, resulting in further warming of waters and disruption of typical weather patterns and ocean circulation. Some believe that a warming Arctic threatens national security (Strawa et al., 2020) and may threaten local and global health by expanding areas of disease vector habitat and releasing viruses that have long been encased in permafrost (National Academies of Sciences, Engineering, and Medicine et al., 2019).

A challenge many educators face is getting their students to be aware of and care about the Arctic when most of us do not live in or near this region. As an educator in northern California, the author well understands this challenge.

This chapter presents lessons and strategies the author successfully used to teach middle-level (11–14-year old) environmental science students in northern California. Despite our distance from the Arctic, by the end of the learning unit, students were able to identify how our actions affect larger environmental systems and how changes in the Arctic influence systemic changes in our area.

DOI: 10.4324/9781003486961-4

Understanding Place-Based Education

In order for students to be concerned about global environmental issues, they first need to be aware of and involved with local concerns. This may be done by anchoring lessons in a place-based approach. With place-based learning, lessons are centered in local context. Rather than relying solely on a textbook for content, teachers may relate learning to the local community through use of outdoor lessons and observations, field studies, and use of local and regional documents and data.

Place-based education (PBE) connects students with their community and the world around them. When taught in the context of a known place, lessons become personal and relevant. Learning may then be scaffolded from local to regional, and then global applications. Making local observations year after year builds a database that students may use to discover trends and anomalies, identify local issues, suggest solutions to local problems, and create reports using data-based documentation.

Gregory Smith Writes

Place-based education is an approach to curriculum development and instruction that directs students' attention to local culture, phenomena, and issues as the basis for at least some of the learning they encounter in school. It is also referred to as place- and community-based education or place-conscious learning. In addition to preparing students academically, teachers who adopt this approach present learning as intimately tied to environmental stewardship and community development, two central concerns of Education for Sustainability. They aim to cultivate in the young the desire and ability to become involved citizens committed to enhancing the welfare of both the human and more-than-human communities of which they are a part. At the heart of place-based education is the belief that children of any age are capable of making significant contributions to the lives of others, and that as they do so, their desire to learn and belief in their own capacity to be change agents increase. When place-based education is effectively implemented, both students and communities benefit, and their teachers often encounter a renewed sense of professional and civic satisfaction.

(Smith, 2017)

PBE provides an opportunity for classes to utilize Indigenous and local knowledge. Inclusion of Indigenous knowledge engages underserved and under-represented groups such as Alaska Native and/or Native American students. The Indigenous knowledge that students learn from their Elders and/or other community experts will provide scaffolding for learning new things and constructing new understanding (Sparrow, personal communication, 25 November 2022).

When updating the California science framework to align with the national Next Generation Science Standards (NGSS), the framework committee recognized that it is important to teach science content in context of the natural world (ideally, the local natural world). The California Environmental Principles and Concepts (EPC) were added to California's science framework in addition to NGSS.

The five Environmental Principles and Concepts are the following.

- People depend on natural systems.
- People influence natural systems.
- Natural systems change in ways that people can benefit from and can influence.
- There are no permanent or impermeable boundaries that prevent matter from flowing between systems.
- Decisions affecting resources and natural systems are complex and involve many factors.

California science teachers of years K–12 (kindergarten through senior year of high school) are to use lessons that support environmental literacy and highlight the deep relationship between humans and the natural world. When planning instruction, all science teachers can keep the EPC in mind as a way to frame lessons. Environmental literacy builds critical thinking and analytic skills that prepare students to make informed decisions about issues affecting our world, and readies them for college and career success (California Environment and Education Initiative, 2021). Using local or regional backdrops for lessons gives the learning most meaning. Environmental literacy builds critical thinking and analytic skills that prepare students to make informed decisions about issues affecting our world, and readies them for college and career success.

Place-Based Education Empowers Students

As students learn through PBE, they become aware of the characteristics of the natural world around them. Students may notice issues that need to be addressed, and then apply their knowledge to create stewardship projects important to their own communities. Young students may simply share a lesson with someone at home, increasing awareness. Other students might develop a school or community garden, conduct audits of energy use, or create educational campaigns about local habitats. These projects may be a short assignment or a course's culminating assignment. Action-oriented projects demonstrate the relevance of learning experiences, give students a personal connection to both learning and their community, and allow students to voice and take responsibility for potential solutions. Through raised awareness of environmental issues and the opportunity to enact change, youth have the power to make a difference.

Examples of Place-Based Programs

San José State University (San José, California) offers an undergraduate course on climate change taught over one academic year. The goal of the course is to promote lasting responsible environmental behavior, and includes student-designed community action projects. When surveyed at least five years after taking the course, participants reported that they continue to make choices to reduce their carbon footprint and environmental impacts (Cordero et al., 2020). The program's lead investigator believes that the local context and action-oriented approach are critical parts of the course's success in creating long-term behavioral changes (Cordero, personal communication, 30 October 2022).

The GLOBE Program

The Global Learning and Observations that Benefit the Environment (GLOBE) program invites students and community members to participate in observations, field studies, and data collection activities. Over 38,000 students in 127 countries participate in GLOBE.

The GLOBE website (globe.gov in the United States) offers an extensive collection of lessons, activities, and research protocols. Lessons are organized by Earth systems – atmosphere, biosphere, Earth as a system, hydrosphere, and pedosphere – and may be searched by grade level. Some lessons are specific to the Arctic. For those living elsewhere, Arctic studies provide useful data to use for comparison of when, for example, bird migration takes place in the Arctic versus locally.

A strength of GLOBE is that participants act as scientists do when conducting research. Participants follow protocols for data collection, collaborate with others, and communicate findings. Following established protocols improves students' science skills, builds confidence in the data, allows for replication, and opens communication among global participants. Students may not be able to read a project's comments posted in Latvian without using a translation app (application) – but they will know how the data was collected and what it was measuring, because protocols were used.

When asked to explain why GLOBE's approach to learning is so effective, the Alaska GLOBE Partnership Director Elena Bautista Sparrow wrote GLOBE's approach to learning about the local environment and impacts of climate change is effective because it is placed-based and also uses a constructivist pedagogy that allows learners to develop their own meaning as they connect the new information presented to their existing knowledge. This learning by doing is accomplished through students actively engaged in observing, measuring, making sense of their data, communicating their findings and applying their new knowledge to a stewardship project related to climate change in their communities (Sparrow et al., 2021). It also allows the students to connect globally through ongoing investigations all over the world by students like them and by scientists using GLOBE data (Sparrow et al., 2013). This process of teaching and learning with GLOBE, braided with Indigenous knowledge and ways of knowing, began in 2000 through the Global Change Education Using Western Science and Native Observations Program (which I led and was funded by NSF), also called the Observing Locally, Connecting Globally (OLCG) program (Gordon et al., 2005; Sparrow et al., 2006), and currently in a NASA-funded project (which I lead): Arctic and Earth STEM Integration of GLOBE and NASA assets (SIGNs), which works with educators and community members in engaging youth in STEM (science, technology, engineering, and mathematics) and climate learning in Alaska and beyond (Sparrow et al., 2019; Spellman et al., 2018). The Arctic and Earth SIGNs learning and inquiry cyclic framework is based on the knowledge of Elders and other local experts, GLOBE, and other NASA (the U.S. National Aeronautics and Space Administration) resources. Also, as part of this project, a curriculum rooted in Dene Athabascan values, tradition, language, and stories was expanded to include GLOBE and NASA resources and other traditional knowledge (Dibert et al., 2021). There is increasing interest in monitoring the environment using GLOBE, as part of climate adaptation planning by Alaskan Indigenous communities (Dibert et al., 2021).

The GLOBE Observer app allows for easy data collection in the field by students and their communities. At the time of this writing, participants may record observations for projects studying cloud coverage, mosquito habitat, land cover, and tree coverage. Observations strengthen some NASA projects. Satellites provide a top-down view of clouds and landscapes; GLOBE observations provide a view from the ground up.

Participation in GLOBE and access to activities are free of charge. Participants must undergo training to be able to upload data to the GLOBE website and access data from other sites. Much of the training is online. The app includes training. GLOBE is a useful tool to use for both school and community citizen science projects. Students participating in science fairs may consider using GLOBE data as part of their work.

Examples of Lessons Using Place-Based Education

The place-based science lessons described here offer opportunities for students to practice important science skills – observation, measurement, documentation, communication of findings, and application of learning to real world problems.

The following lessons describe how these skills are incorporated into activities.

Studying Biotic and Abiotic Factors

Lessons and lectures may easily transition to a PBE approach. For example, rather than just giving students the definition of *biotic* and *abiotic* factors, take the class outside, or at least to a window. Ask students to photograph or sketch the scene, and then identify what biotic and abiotic factors are found in the school environment. Learning the new terms in the context of the school campus helps students attach more meaning to the concepts. This simple activity can be returned to throughout the school year, noticing how changes in seasons and possible changes in infrastructure affect local biotic and abiotic factors.

Phenology Studies

To accurately describe environmental changes that are happening – likely because of climate change – baseline conditions need to be documented. When average temperatures, precipitation, plant growth patterns, and presence of wildlife are known, deviations from a typical measurement may then be noted.

Phenology is the study of the natural cycles of plants and animals. Natural cycles include the timing of bird migration and nest building, insect hatching, and appearance of plant flowers and leaf buds. These simple events are important ecological markers. A change in timing of one event – hatching of insects, for example – may affect an ecosystem and the food webs within it (Brinker, 2013). According to the National Phenology Network, "Changes in phenological events like flowering and animal migration are among the most sensitive biological responses to climate change" (USA National Phenology Network, 2022).

Phenology studies may lead to student research projects, research collaborations with classes from other locations, or participation in citizen science projects. Using phenology studies helps students hone their qualitative and quantitative observational skills (Herrmann, personal communication, 22 November 2022).

A side benefit from conducting phenology studies is positive mental and physical health outcomes for both students and adults. Mental and physical health benefits are associated with more time spent outside in nature (Beyer et al., 2018). Students in urban settings may spend less time outdoors than their suburban and rural peers, and will benefit from nature studies. Studies show that youth and adults spend more hours per day using electronics than being outside, connecting to nature (Yale Environment 360, 2017). Environmental studies likely utilize electronics for documentation and measurements. Teachers may build in device-free time when outdoors. Include times of silence and observation, use of binoculars, and sketching to encourage students to pause and more deeply connect with the world around them.

Tree Phenology

Setting Up the Study

Select a tree(s) or bush(es) that are in a safe location, are easily accessible, and are in a place where students will be within view of the teacher while they work. Ideally, the plant should be a native species that is not on a managed landscape. If a native plant is not available to study, do not let that stop the investigation. Those living in areas with no trees may use wildflowers, bushes, or grasses for the study. Studying multiple plants in the area will allow teams of students to work at the same time.

Use an aerial map of the study area and assign each plant a number. Labeling and documenting each plant is helpful for record keeping from year to year. Also, if this project is passed on to a new teacher to continue, he or she will be able to confidently identify which plants are in the project.

Many scientists use notebooks in the field. So should students. Assign a bound composition notebook to each plant. Student teams record observations in each notebook. The notebooks may be used from year to year.

In a school year, make the first observation at the start of the year, and then approximately every six weeks. Collect the following data.

The observation's time and date, including year.

Circumference of the tree trunks, if observing trees. Using measuring tape, students measure the circumference 1 meter up from the ground. Plant height may be measured. The purpose of this measurement is to track growth. This data may also be used to calculate the amount of carbon stored in the tree (SERC, 2021).

Color of the leaves, if leaves are present, using a GLOBE Plant Color Guide (Figure 4.1). Place the color guide over a leaf and select a color that most closely matches the color showing on at least half of the leaf. Each color has a standardized shade and number. When the color number is reported in data, others know if the leaves were light green, dark green, yellow, or red.

Description of the plant and leaves, written in the students' own words. The description may include leaf color; any color changes noted; how much of a tree canopy has leaves; presence of flower buds, flowers, or leaf buds; and any other observations made. Sketches or photographs of the plants may be included.

Date of bud break. Bud break is when a plant's leaf or flower buds start to open; bud break date is an important phenologic marker. These dates are watched by fruit, nut, and grape farmers and mark the start of the growing season. Bud break is occurring earlier in some areas. Track when bud break occurs in your area.

Date of green down. Green down is when leaves change color from green to autumn colors. This marks the end of the plant's growing season. By tracking bud break and green down dates for the same plant, students will observe the plant's productive growth cycle. Chart these year to year and observe for shifts in dates. Protocols for taking bud break and green down observations may be found on the GLOBE website.

Student signatures. By signing their work, students take ownership of the effort. When students know that the notebook will be used by their peers in the following years, they will likely take more effort with their work (Figure 4.2).

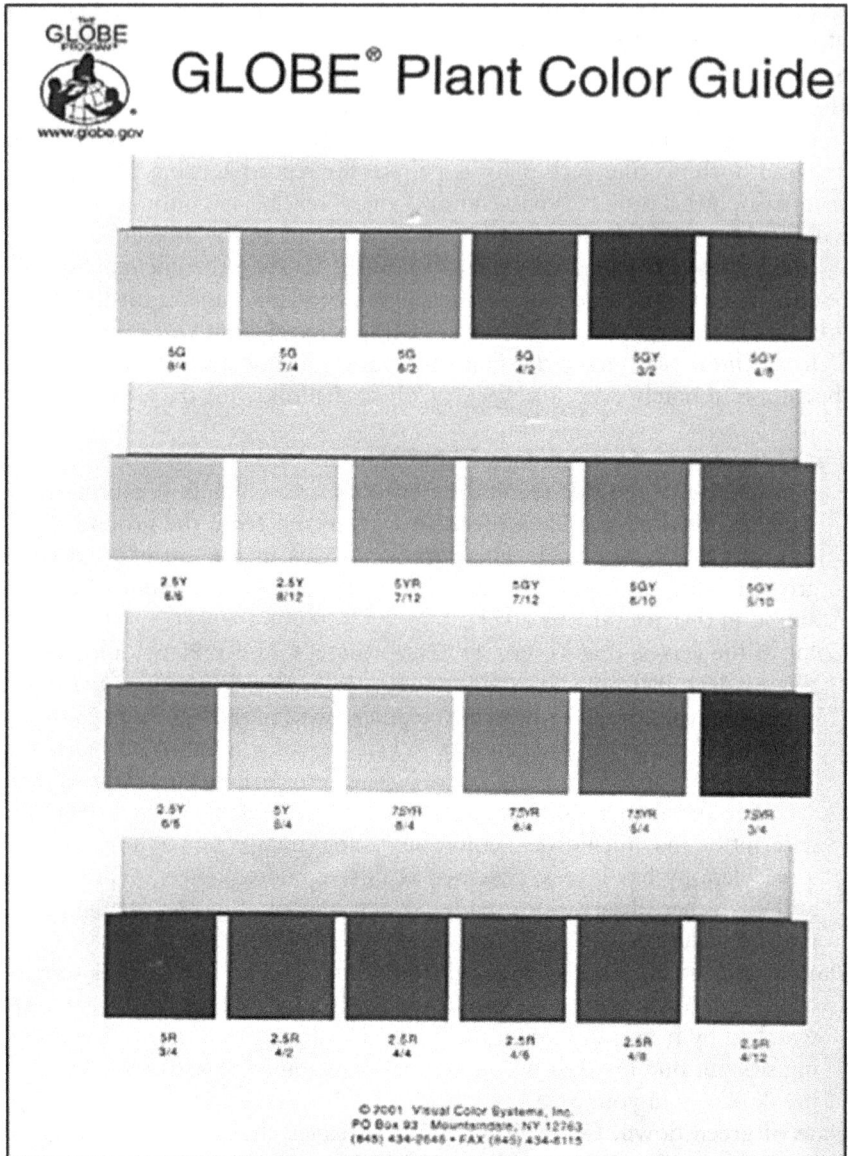

Figure 4.1 GLOBE Plant Color Guide

   Observations are made approximately every six weeks, depending on the plants' status. Keep watch on the plants. When leaf colors start to change, leaves begin to drop, buds and/or flowers appear, and flowers and leaves open, take the students outside for observations. When possible, make an observation on the day a new season starts.

Figure 4.2 Students documenting blossoming of ornamental cherry trees on a school campus

Before going outside, students should first practice procedures in the classroom, where attention is more focused. Before each outdoor observation, review measurement and documentation protocols, reinforce behavior expectations, and remind students what they should be looking for.

Phenology studies may be interwoven into science lessons about earth systems, earth's seasons, weather, climate, ecosystems, food webs, and climate change.

Field Studies

Water quality studies provide another opportunity for place-based learning. Water quality parameters include water temperature, pH, dissolved solids, dissolved oxygen, and turbidity. Students may also note and describe litter found in or near the water. GLOBE offers protocols for water quality assessments. Adding phenology and weather measurements to water studies broadens the scope of a field study. These can include air temperature, cloud cover, recent precipitation, presence of trees and plants and their leaf color, and what wildlife are seen. Students become more aware of the relationships between aquatic and riparian systems with this field study.

When repeated water quality assessments are done, have students create a graph for each of the quantitative measurements. Graphing the data will help students analyze results and visualize trends, especially when graphs are compared from season to season and year to year.

Students who do not have access to a body of water may partner with a school that directly assesses water quality and interpret the data supplied by other students.

Over the seasons, aquatic areas change. Rain prompts new plant growth. Fish, birds, insects, and other fauna respond to water levels, drought, and changing seasons. It is easy to take seasonal changes for granted and overlook what is happening in nature. Field studies prompt students to slow down, observe, and reflect on natural systems.

Monitor these changes by photographing the field site each visit from marked positions. These photos may be used in the classroom for further discussion and evaluation. Students may use the field site photographs and data and create food web diagrams. The field site photographs may also be used in discussions of habitats, weather, and climate change. Students may

Figure 4.3a–c Field study photos over the seasons
Source: Brinker, 2021

Figure 4.3a–c (Continued)

observe and discuss the changes to biotic and abiotic factors in the field site in response to a change of seasons, a severe storm, drought, pollution, and actual or theoretical habitat encroachment.

Once students are familiar with seasonal changes in their field study location, expand the discussion to larger areas. After record rains occur in a different location, ask students to explain how their field study area would be affected by such a storm. Connect these changes to the Arctic. As permafrost thaws, how are local ecosystems impacted? As polar glaciers melt and sea levels rise, how will aquatic and riparian ecosystems be affected? Using a local setting as a reference for impacts of climate change, students will better appreciate potential consequences of environmental change in other regions.

Ongoing field studies provide an opportunity for multi-class collaborations. Students may compare the photos, observations, and data collected by peers in other regions or countries with their own data. Sharing this information will personalize science lessons about biomes, ecosystems, and climate change.

Field studies may lead to collaborations with local agencies. One high school science class regularly monitors a local creek's water quality and documents the presence of fish, frogs, insects, and other markers of a location's environmental health. Because of their field experience, the students became part of a local watershed restoration program. Students work with scientists and learn about career and internship opportunities. Field work has inspired several students to major in ecology or environmental science in university.

By noticing when natural cycles occur and if this timing changes, students witness examples of the interconnectedness of life on Earth. Tracking anomalies in local weather events – a storm bringing higher than average precipitation, a heat wave, drought – and any impacts on phenology markers could be a predictor of future conditions. If leaves of fruit trees are staying green later into the growing season will this affect subsequent fruit production? These anomalies may become new climate patterns.

Conduct a Bioblitz

A bioblitz (Figure 4.4) is a communal citizen science effort to record as many species as possible. A bioblitz's area of study may be small or large. It is the organizer's choice. A bioblitz may focus on one topic – bees or plants, for example – or include any living thing found in the study area. A bioblitz may last for a day or a week. Again, it is the organizer's choice. A bioblitz encourages collaboration. The event may be organized to include observations from students, families, and communities.

The iNaturalist online community supports both bioblitz activities and one-time observations. Users may use the iNaturalist app to upload individual observations, create a project, or contribute to posted research projects.

Figure 4.4 University and elementary level students working together to make observations for a bioblitz.

Source: Brinker, 2011

Findings are shared with the Global Biodiversity Information Facility and other scientists.

To organize a bioblitz:

- Identify the geographic area for the study. If necessary, mark this area on a map to share with participants.
- Identify date(s) and time(s) of the bioblitz. The activity might span one class session or a full 24-hour day.

- Participants download the iNaturalist app. Use the app to photograph observations, pinpoint location, and share observations with the project. If possible, participants identify what flora or fauna is observed. If they are unsure, others will identify the subject based on the uploaded photograph. The iNaturalist website has online field guides related to specific project areas or topics. At the time of this writing, more than 400,000 species have been observed using iNaturalist (iNaturalist, 2022).

Finding Connections

At the end of the 12-week environmental science class (Figure 4.5), students were asked to demonstrate ways their activities and lessons connected to the larger issue of climate change.

Figure 4.5 Students on a BioBlitz

First, we reviewed what the class learned and accomplished. We created a master list of topics studied and associated projects. Students then joined with a partner and identified key points related to each topic. Students wrote each topic from the master list onto sticky notes (Figure 4.6). *Climate change* was written in the center or across the top of a large piece of paper.

Each team discussed what its members learned about the master list topics, and how these were connected. Class notes could be used as reference. During their discussions, students moved the sticky notes around the chart, creating a concept map. Students next demonstrated their learning by adding

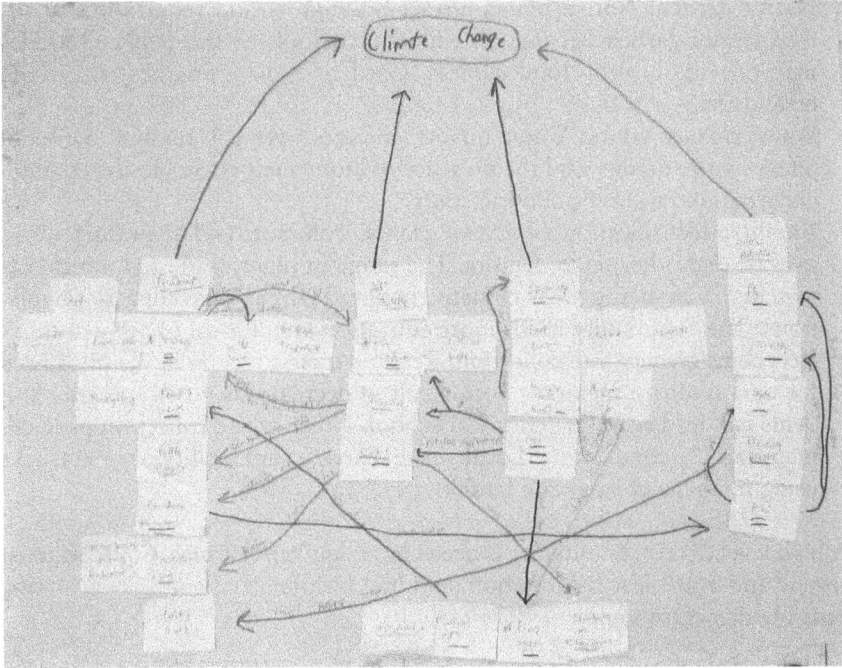

Figure 4.6 Student work in progress

supporting facts and details under the topic headings. The last task was to draw arrows connecting the topics together and write, on top of the arrows, a brief explanation of the connection. The arrows might lead to and from one topic. Several topics may be connected. All should, in some way, link back to climate change in a positive or negative way. Two class periods were given for the full activity.

Following is an example of study topics and key points.

- **Climate change:** Evidence supports climate change caused by human activity. Climate change effects are felt locally and globally.
- **Sources of energy:** Pros and cons of fossil fuels and renewable energies. Burning of fossil fuels emits CO_2, traps atmospheric heat, and increases air pollution. Fossil fuels are used to create plastics, including single-use items. Renewable energy sources do not emit CO_2.
- **Air quality assessment:** Vehicles and wood burning produce gasses and particulates that contribute to poor air quality. High particulate matter and pollen levels in the air can lead to health problems. Vehicles may idle in the school parking lot, increasing student exposure to emissions. Trees store CO_2.

- **Native garden:** Native plants attract pollinators and require less water. The school garden became an educational tool for the public. Gardens may provide healthy food choices. Gardens radiate less heat than concrete does.
- **Water quality study:** Water quality changes over the seasons. Drought affects water quality and the presence of biotic factors. Single-use containers are found to be a pollution source.
- **Bioblitz:** Identification of native plants, common and migrating birds, insects, and other biotic factors. The range of plant, bird, and insect ecosystems is changing with climate change. Habitat loss impacts populations. The community needs more education about local native plants.
- **Recycling/green waste collection:** Reduces trash to landfills. Reduces need for new resource material. Both result in decreases in fossil fuel use. Students can lead recycling efforts at school. Green waste collection provides materials for composting. Local waste management facility powers trucks using methane created at a landfill site.

A high level of critical thinking is needed to complete this task. Students must review and synthesize information, and find meaning and connection among possibly disparate topics.

Summary

The Arctic is warming up to four times faster than the rest of the world. Changes happening in the Arctic affect us all. It is important that we understand the implications of a warming Arctic, regardless of where we live. Educators in nonpolar regions may find it challenging to have students find relevance in polar science. Place-based education (PBE) is an effective instructional strategy to use to increase students' understanding of the impacts of climate change.

PBE personalizes learning through use of local settings as context for instruction. Students may then apply their knowledge to regional and global scenarios. When students understand the impacts of a warming climate on local water sources, for example, they are likely to describe the impacts of drought on other communities. PBE supports use of field studies, citizen science projects, collaborative research, and student action projects.

References

Beyer, K, Szabo, A, Hoormann, K, and Stolley, M. 2018. 'Time spent outdoors, activity levels, and chronic disease among American adults', *Journal of Behavioral Medicine*, Vol. 41, No. 4, 494–503.

Brinker, R. 2013. 'Phenology and nature's shifting rhythms', *Ted-Ed*, viewed 26 July 2022.

California Environment and Education Initiative. 2021. 'California's environmental principles and concepts', *California Environment and Education Initiative*, viewed 28 July 2022.

Cordero, EC, Centeno, D, and Todd, AM. 2020. 'The role of climate change education on individual lifetime carbon emissions', *PLoS ONE*, Vol. 5, No. 2, e0206266, viewed 20 October 2022.

Dibert, M, Sparrow, E, Chase, M, Buffington, C, Spellman, K, and James, N. 2021. 'Learning from K'keeyh: Connecting generations and multiple knowledge systems through cross-cultural learning and teaching', Abstract Number ED52A-02 presented at the American Geophysical Union Meeting 13–17, December.

Gordon, L, Stephens, S, and Sparrow, EB. 2005. 'Applying the national science education standards in Alaska: Weaving native knowledge into teaching and learning environmental science through inquiry', in Yager RE (ed), *Exemplary Science: Best Practices in Professional Development*. NSTA Press, Arlington, pp. 85–98.

iNaturalist, 'Overview, projects, guides', viewed 26 September 2022.

Jacobs, P, Lenssen, N, Schmidt, G, and Rohde, R. 2021. 'The Arctic is now warming four times as fast as the rest of the globe', *Proceedings of the American Geophysical Union, New Orleans*, viewed 20 August 2022.

National Academies of Sciences, Engineering, and Medicine; InterAcademy Partnership; European Academies Science Advisory Council; Division on Earth and Life Studies; Health and Medicine Division; Polar Research Board; Board on Life Sciences; Board on Global Health; Lauren Everett, Rapporteur. 2019. 'Understanding and responding to global health risks from microbial threats in the Arctic', *National Academy of Sciences, Engineering, and Medicine, Proceedings,* viewed 20 August 2022.

SERC. 2021. *Living in a Carbon World; Carbon Storage in Local Trees*. Science Education Research Center at Carleton College, viewed 24 November 2022.

Smith, G. 2017. 'Place based education', *Oxford University Press,* viewed 10 August 2022.

Sparrow, E, Spellman, K, Chase, M., Buffington, C., Murray, B., Yoshikawa, K, and Sparrow, EB. 2019. 'Personal and cultural connections: Key to science communication, education and engagement (invited)', Abstract Number ED12B-01 presented at 2019 Fall Meeting, AGU, San Francisco, CA, 9–13 December.

Sparrow, EB, Gordon, LS, Kopplin, MR, Boger, R, Yule, S, Morris, K, Jaroensutasinee, K, Jaroensutasinee, M, and Yoshikawa, K. 2013. 'Integrating geoscience research in primary and secondary education', in Tong V (ed), *Geoscience: Research–Enhanced School and Public Outreach*. Innovations in Science and Technology 21. Springer, London, pp. 227–250. DOI: 10.1007/978-94-007-6943-4.

Sparrow, EB, Stephens, S, Gordon, LS, and Kopplin, MR. 2006. 'Weaving native knowledge and best teaching practices in GLOBE implementation', *Proceedings of the Tenth Annual GLOBE Conference*, held in Phuket, Thailand. Global Learning and Observations to Benefit the Environment Program.

Spellman, KV, Sparrow, EB, Chase, MJ, Larson, A, and Kealy, K. 2018. 'Connected climate change learning through citizen science: An assessment of priorities and needs of formal and informal educators and community members in Alaska', *Connected Science Learning*, Vol. 1, No. 6, 1–24.

Strawa, A, Latshaw, G, Farkas, S, Russell, P, and Zornetzer, S. 2020. 'Arctic ice loss threatens national security: A path forward', *Orbis*, Vol. 64, No. 4, 622–636.

USA National Phenology Network, *Why Phenology?* USA National Phenology Network, viewed 10 October 2022.

Yale Environment 360. 2017. 'U.S. study shows widening disconnect with nature, and potential solutions', *E360 Digest*, Vol. 27, April 2017.

5 Sustainable Education for the Youngest

The Implementation of New Technologies to Foster Awareness of Environmental Problems in Elementary and Secondary Schools

Inga Beck

Introduction

It is no new wisdom that sustainable education has to start as early as possible in order to create a society that takes care about its environment and internalize a sustainable way of living in daily life.

Fortunately, it is the youngest generation who are open to new things and who still have very open minds. Children at elementary and secondary school are very eager to learn and are excited about new things. Unfortunately, many recent teaching practices are not meeting the entire needs of this age group, and the potential that is slumbering in each of them often hence is not awakened. Kids want to learn while still playing and by exploring by themselves. Furthermore, they are very interested in new technologies, they love using smartphones and tablets, and although there might be many objections for the abuse of technical devices, there is no doubt that a responsible use of it has to be taught. Many innovative tools and methods are freely available, and it is our task as educators to say goodbye to old-fashioned teaching methods and to dare to approach and use them. Thereby, the following five important pedagogic goals will be reached.

1. Child-friendly teaching material is used that
2. promotes individual skills;
3. the (responsible) handling of electronic devices, which fosters the children's independency;
4. guarantees education for sustainable development; and
5. as these methods often are conceptualized for small groups, social abilities are promoted.

This chapter introduces the so-called Action Bound, one such a state-of-the-art tool that combines all the previously mentioned necessities for learning while playing and exploring by using new technology alone or in a group. Additionally it is even applicable outdoors and can be combined with creative or physical activities. Three different applications – all under the umbrella of environmental education – will be presented and a how-to-use it in your own

DOI: 10.4324/9781003486961-5

classroom guide is provided. The program quite literally augments reality by enhancing peoples' real-life interaction while using their smartphones and tablets. It is therefore possible to create an application-based do-it-yourself (DIY) escape game, a digital timeline of events or places of interest tour, with the use of GPS coordinates and pre-placed codes and mysteries.

Introduction of Action Bound

The following subsections will highlight three different Action Bounds, with three different intentions and three different target groups. All were created in the context of environmental education. As they should be used within different settings, methods and tools used within the Bounds vary from one Bound to the other.

Short History and Background

Action Bound started in Berlin, Germany in 2012, and it developed enormously since then. It started out as an educational project, but it is now used by people all over the world for all kinds of different things.

Since then, it received awards such as the 'eLearning Award', the 'Pädagogischer Medienpreis' (educational media award) and the 'EduMedia' seal (Figure 5.1).

Figure 5.1 Awards received by Action Bound
Source: Action Bound (2022)

Target Group and Implementation

Action Bound is an app for playing digitally interactive scavenger hunts to lead the learner on a path of discovery. These multimedia-based hunts are called 'Bounds'. For teachers or educators, it offers a new learning-platform in three different ways – they can either use exiting Bounds that are dealing with a topic that fits into the learning plan (you can find a list of Bounds at the end of this chapter), they can create their own Bounds tackling specific questions or issues that they want the children to learn about, or they can let the pupils create their own Bound on a certain topic.

The Bounds can be played on either a smartphone or tablet. The only requirement is the installation of the Action Bound app, which is available for all common gaming systems. The app is free of charge and can be found for iOS, Android and PC formats. Every Bound comes with a specific quick reference (QR) code which has to be scanned with the app. After scanning, the Bounds opens and guides you through the process. It is now possible to download the information of a Bound and play the Bound later offline. The Bounds are from different providers: private persons, organizations, museums, schools, etc. It depends on the provider if the Bound is free of charge or not.

For the creation of a Bound, an account at the website https://en.actionbound.com/?setlang is necessary. If the account is only used for private purposes, there are no costs. For schools or other educational organizations, a registration is necessary, as well as an annual fee.

With an account, it is now possible to create a Bound no matter which topic. The program offers a huge potential of gamification with the Bound Creator's extensive game elements and tools like GPS locations, directions, maps, compass, pictures, videos, quizzes, missions, tournaments, QR codes and much more to create fun and exciting mobile app-based adventures (Figure 5.2).

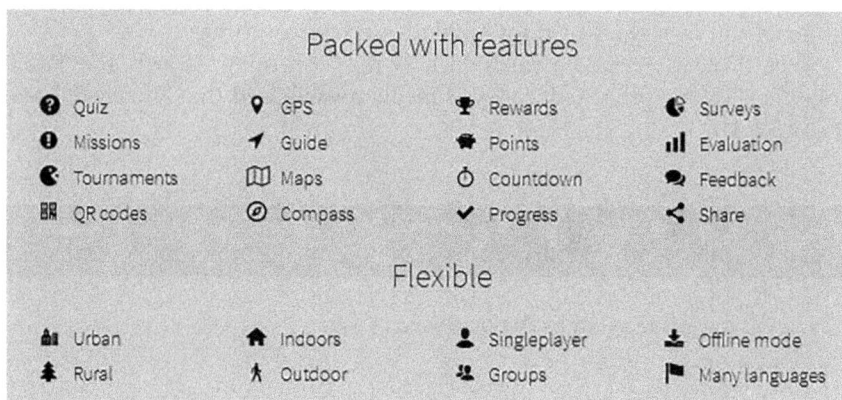

Figure 5.2 Features provided by Action Bound
Source: Action Bound (2022)

The following case studies will highlight three different Action Bounds, with three different intentions and three different target groups, but they were all created in the context of environmental education. As they should be used within different settings, methods and tools used within the Bounds vary from one Bound to the other.

Case Studies

The three case studies presented here were, as mentioned, all created in the context of environmental education with the overall goal to foster the awareness about the nature. However, they differ in many ways.

'Hilf Gletschi Beim Überleben' Action Bound About Climate Change for
the Environmental Research Station Schneefernerhaus

In 2021, the Environmental Research Station Schneefernerhaus (UFS GmbH) was involved in a Bavarian/Austrian project called 'KlimaAlps'. The UFS GmbH is Germany's highest research station, located just below the summit of Zugspitze (2,962 m). Many different institutions use this building as platform for high-end scientific research with the main focus on climate change and its consequences for humans and the ecosystem.

The project KlimaAlps (www.klimaalps.eu) was a joint initiative of six partners from Bavaria and Austria led by the German 'Energiewende Oberland'. The mission of the project was to make the impacts of the climate change that are happening in the nature visible to everybody in order to create more awareness for environmental issues within the population.

Within this project, the UFS GmbH created an Action Bound that should give everybody insights about the building and the research conducted there. Hence, the Bound was created for individuals and can be played anywhere and at any time.

The story behind the Bound is the mystical creature 'Gletschi' (Gletscher is German and means 'glacier') who lives close to the Schneefernerhaus. It depends on clear snow and ice from the glacier to survive. As climate change happens, it realizes that it also loses its nutritional basis. Hence, it looks for help to save its glacier – and hence, the environment. The player is now requested to help him by answering different tasks in the context of nature conservation. For every correct accomplished task, Gletschi shows him some insights of the Schneefernerhaus.

For the tasks, it was taken into account that the most different methods as possible available at Action Bound were embedded, in order to keep the 'play' interesting (e.g., there are videos showing wrong environmental behavior which the player has to identify, there are quizzes about finding the most sustainable food, etc.).

Also, the 'rewards' vary. As Gletschi shows a video of a part of the Schneefernerhaus, scientists shortly explain their research or photographs give the player impressions about the building.

This Bound was awarded by Action Bound and it was presented as a Showcase during the tenth anniversary celebration of Action Bound. A video of this presentation is available at the Action Bound YouTube channel by searching 'Gletschi – Showcase'.

Action Bound About 'Raus in die Natur' for the Federal Nature
Conservation

The second and the third Bounds presented as case studies were both create by the federal nature conservation in Bavaria for the location Wartaweil (https://www.bund-naturschutz.de/umweltbildung/bildungsstaetten/wartaweil). 'Raus in die Natur' ('Into the Nature') is a Bound fashioned for families

Figure 5.3 Layout Action Bound 'Hilf Gletschi beim Überleben'

Figure 5.4 QR code to get access to the Bound

who would like to get outside while playing, moving and learning something about nature. It has to be played outside in the area of Wartaweil. As there is no internet connection, the Bound must be downloaded beforehand.

A beaver guides the players through the area. They are prompted to find special plants or animals, count flowers and take pictures of them or to perform certain little physical tasks. While playing, information about the forest and the biodiversity is given in the form of little texts, videos or images.

This Bound uses a very useful feature offered by the program: Depending on the season, the player partially gets different questions and information. Hence, during springtime, the focus of the Bound is on the offspring, in summertime on the water cycle in the forest, in autumn on the foliage coloring and in winter on hibernation.

The intention of this Bound is to get families out into nature and to playfully become educated about the forest. It can be played anytime and is free of charge.

Action Bound About 'Biodiversität' for the Federal Nature Conservation

For a more educational purpose on biodiversity, the Biodiversität Bound was explicitly configurated for school classes. As for the 'Raus in die Natur' Bound, it is necessary to download the Bound first and then go out and play. In contrast to the first two Bounds presented, this Bound is conceptualized for groups. Hence, teachers can come with their classes, divide the class into groups and let the students go with their tablets to play. The target group is children between 10 and 13 years, and the content of the Bound goes along with the learning plan predetermined by the Bavarian State Ministry for Education.

The special thing about this Bound is that additional material is handed out by the federal nature conservation Wartaweil. For example, ropes are needed to create a food web during playing and tree slices to learn about the meaning of the different tree rings.

With this Bound, the teacher can easily let the children play while they are exploring nature by themselves and get educated by the information provided by the Bound. The teacher's role is to guide the groups through the process and help out if questions arise. Afterwards, is it expected that the educators discuss the content and the learned within the entire class.

How-to Guide

The first step to make an Action Bound is the definition of the content, aim and target group of the Bound. Afterwards, it is very useful to draw the ideas in some kind of a flowchart and to create a common theme.

Once registered and logged in at Action Bound, a window opens showing to create a new Bound (Figure 5.5; Point 1 and 2). After the creation of a new project on the left side, the following four options appear (Figure 5.5; Point 3).

- Your Bound
- Content
- Settings
- Results

The first point contains short information about the Bound has to be filled out. This information will be visible for players and give them a first impression about the content of the Bound.

Afterwards, the creation of the Bound can be started under 'Content'. This is very intuitive: In the middle, a little + appears; by clicking on it on the right side, different modules (quiz, question, survey, etc.) can be put together by the creator, edited, exchanged or deleted at any time (Figure 5.5; Point 4).

Once a first framework is finished, further tools can sophisticate the Bound. An upload function offers the possibility to amplify the Bound with pictures, videos or audio.

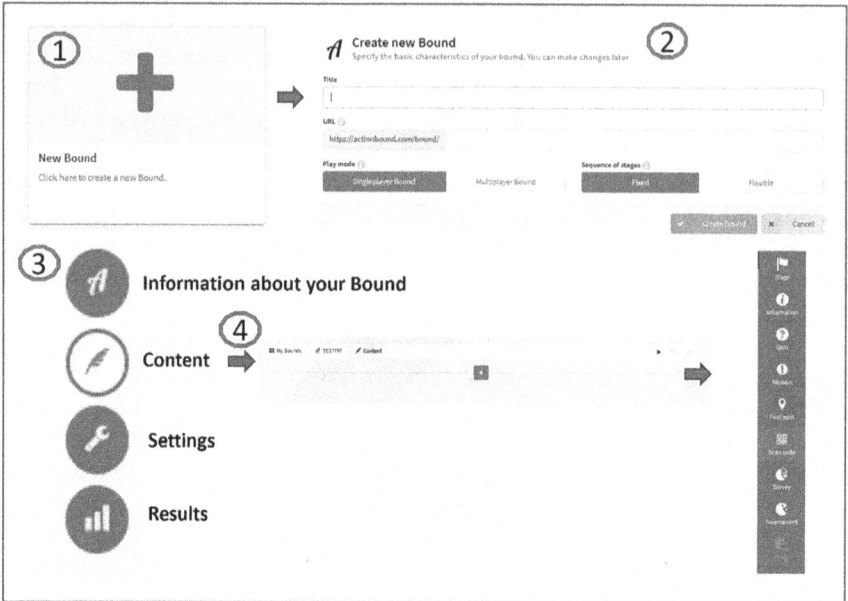

Figure 5.5

After finishing, the Bound can be tested several times until every error is fixed and every 'lose end' could be found and deleted.

There is a very comprehensive 'Beginners Guide' available at:

https://www.youtube.com/watch?v=kXVcRs88zYo

Further Tips and Help

To ensure that a created Action Bound runs smoothly, it is strongly recommended to have it tested several times by independent and individual persons.

To get started with the program or to become better and get more insight information, Action Bound offers a variety of webinars and free online courses. The content varies from introductions for beginners to advanced information webinars for longtime users. Almost all of the courses are recorded and available on the Action Bound YouTube channel. Additionally, there is an online platform that serves as forum for users to exchange their experiences and to discuss problems.

Furthermore, there is email support available. Every question is answered in a very timely manner, and new ideas to improve the program are welcomed.

A newsletter provides the user with new tools, public Action Bounds and upcoming events.

Examples of English Action Bounds

The following are three selected Action Bounds that are public and free to use. Their content is all related to environmental issues or are tackling the climate change topic. There are many more Bounds available and searchable at the Action Bound Site under 'public Bounds'.

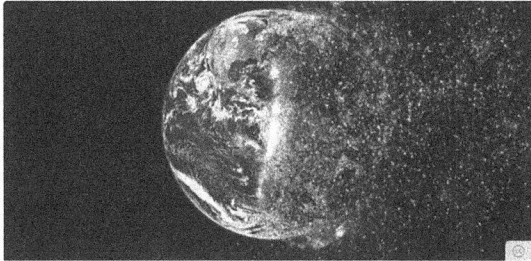

Sie erfahren viel über den Klimawandel, seine Ursachen, Folgen und Lösungen, wie man den Klimawandel bekämpfen kann

Scannen Sie diesen Code mit der Actionbound-App, um den gebundenen Code zu starten.

Figure 5.6 Climate Change and Sustainability (Group Bound)

Scan this Code with the Actionbound App to start the Bound.

Figure 5.7 Nature Wonders (Group Bound)

Heute lernen wir alles über Schmetterlinge in Irland. Hast du gut gelernt?

Scannen Sie diesen Code mit der Actionbound-App, um den gebundenen Code zu starten.

Figure 5.8 Environmental Day (Single Bound)

References

Action Bound. (2021) Beginner's course – Action bound basics. Online at https://www.youtube.com/watch?v=t_sQkCCjm8c (accessed 14 December 2022).

Action Bound. (2022) Mobile multimedia quiz. Online at: https://en.actionbound.com/?setlang (accessed 12 December 2022).

Beck, I. Huffman, L., and Walton, D. (2014) Education and polar research: Bringing polar science into the classroom, *Journal of Geological Resource and Engineering.* 4, 217–221. doi:10.17265/2328–2193/2014.04.004, hdl:10013/epic.45284

Bund Naturschutz Wartaweil. (2022) Digitaler Walderlebnispfad in Wartaweil. Online at: https://www.bund-naturschutz.de/umweltbildung/bildungsstaetten/wartaweil/bildungsangebote/projekte/action-bound (accessed 1 December 2022).

KlimaAlps. (2022) KlimaAlps – Klimawandel sichtbar machen. Online at: https://www.klimaalps.eu/ (accessed 1 December 2022).

May, I., Carlson, D., Ardyna, M., Geoffroy, M., and Heikkilä, M. (2012) Letter-to-the-Editor: Making science animations: New possibilities for making science accessible to the public. *Polar Research.* 30, 15315, DOI: 10.3402, pp.

OHA. (2021) Digitaler Walderlebnispfad. Zeitung aus dem Pfaffenwinkel.

6 Citizen Science and Environmental Education

Case Study of Project "Salvemos el Rio Atoyac"

Gala Perez Gutierrez and
Esther Madrid Morales

Salvemos el Rio Atoyac

The Atoyac River

The name Atoyac derives from the Nahuatl word *atoyatl*, which means "stream of water" or "place of the river" and makes reference to a place that is characterized by the beauty of its river.

The Atoyac River Basin is located in the state of Guerrero, in the so-called region of Costa Grande. It covers an area of 904 km², and partially includes the municipalities of San Miguel Totoloapan, Tecpan de Galeana and Heliodoro Castillo, and mainly Atoyac de Álvarez and Benito Juárez.

The Atoyac River has a length of 74 km; it is born in the northern limit of the municipality of Atoyac de Álvarez and it runs towards the south and crosses the municipality of Benito Juárez to finally flows into the Pacific Ocean (Figure 6.1).

The population that lives in the basin is around 53,000 inhabitants (INEGI, 2015) and is distributed in 116 localities (95 corresponding to the municipality of Atoyac de Álvarez and 21 to Benito Juárez), of which four are considered urban areas, concentrating 67% of the population.

Regarding the land use of the basin, 13% is intended for agriculture (corn, coffee, mango and coconut) (SAGARPA, 2016); another 13% corresponds to grassland area with the development of livestock activities of cattle, goats and sheep; 20% is occupied by the jungle; about 50% corresponds to forest; and the rest comprises urban areas (IMTA, 2018).

Other economic services obtained from the basin are logging and small-scale fishing for crustaceans (*Macrobrachium americanum*, *M. occidentale* and *M. tenellum*).

Atoyac River Pollution

Water pollution is one of the most prevalent environmental problems in Mexico. Only in 2019, intestinal infectious diseases due to the intake of

DOI: 10.4324/9781003486961-6

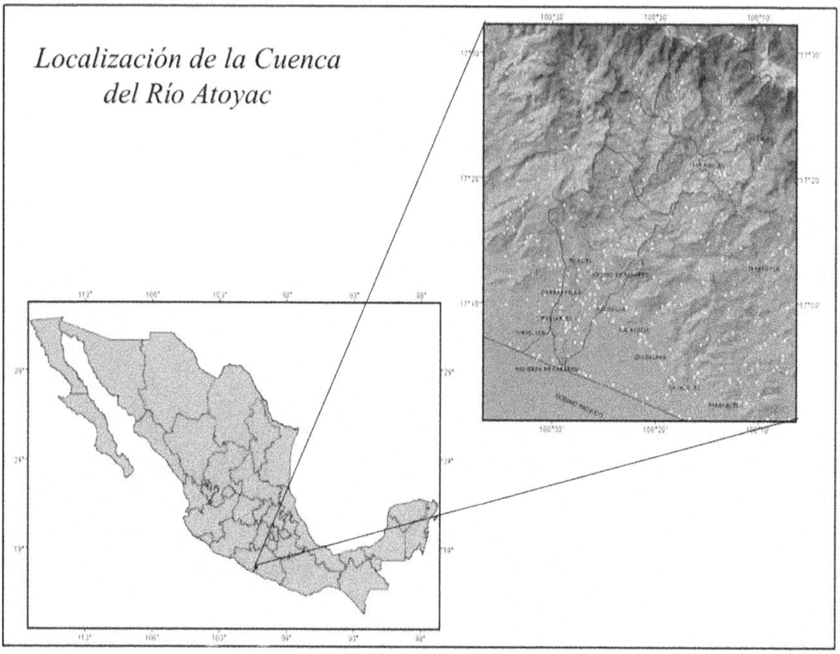

Figure 6.1 Atoyac River location

Figure 6.2 Mountain of Atoyac de Alvarez, Guerrero

contaminated water were the sixth cause of death in children under one year of age, registering 353 deaths.

The problem of contamination of the Atoyac River dates back to the 1990s. In 1996, the Institute of Applied Ecology of Guerrero carried out a study on the control of environmental contamination and the prevention of its effects on human health, and released a report called "Environmental Problems in the State of Guerrero".

The results indicated that the river presented high levels of biological contamination, and at least six discharges of untreated wastewater. It recommended the installation of a treatment plant in order to improve the life quality of the inhabitants by reducing the levels of gastrointestinal diseases, and at the same time avoiding damage to the environment. Since then, numerous environmental impact studies and citizen mobilizations have been carried out to reduce pollution.

Jimenez Varela, et al. (2000) carried out the "Environmental Diagnosis of the Atoyac River Basin", highlighting the poor management of the river. They pointed to the direct discharge of wastewater on the hydrological basin, the direct and indirect dumping of solid waste and illegal logging as the main

Figure 6.3 Clandestine discharge of sewage and garbage

causes of its deterioration. Its results show contaminant levels above the maximum permissible limits according to the Ecological Criteria for Water Quality for different parameters, highlighting the high concentrations of fecal coliform bacteria. Likewise, it was detected that 38% of the surface of the basin was disturbed with land use change from forest and jungle, to agricultural pasture and fragmented forest, favoring soil erosion and the transport of sediments.

In 2007, specialists from the Autonomous University of Mexico (UNAM) carried out the Pro-Regiones project, whose purpose was to promote the sustainable development of hydrological basins in different regions of Mexico. Their results showed a severe impact on the river due to the direct dumping of waste, sewage and agriculture chemical products, causing damage to biodiversity, farmland and human health.

The Municipal Development Plan of Atoyac de Álvarez (2015–2016) confirmed the contamination of the river, adding to the points addressed in other studies, the lack of certification of the municipal slaughterhouse, the deterioration of the sewage system, the malfunctioning of the waste collection and drinking water system, and the existence of clandestine cattle sacrificial canals in some of the main communities such as Paraíso, Ticui, Corral Falso and Ciruelar.

Figure 6.4 Direct discharge of sewage and garbage into the river

Figure 6.5 Deforestation is one of the most important environmental problems in the state of Guerrero

Figure 6.6 Sewage and garbage are one of the biggest problems in the urban area of the Atoyac River

In 2012, the CONAGUA (National Water Commission) indicated in its National Water Information System the presence of high concentration levels of fecal coliforms obtained in the only water quality monitoring station called Puente de San Jerónimo. In the 2017 monitoring, CONAGUA once again measured high contamination by fecal coliforms, highlighting the lack of actions to improve water quality.

Citizen Action and the Creation of "Salvemos El Rio Atoyac"

The citizen movements began at the end of the 1990s, and a "Citizen Council for the Atoyac River Rescue" was formed in 2000. Their role has been crucial in the attending the problematic about the river pollution and their actions can be divided into the following categories.

1. Increasing awareness in the concerned communities.
2. Demanding the authorities take the necessary actions to solve the problems.
3. Educating new generations.
4. Cleaning, restoring and proposing solutions.

Figure 6.7 Constitutive Assembly of the Citizen Council for the rescue of the Atoyac River, August 19, 2003

Among its different activities, the following four river sanitation campaigns carried out in different locations stand out.

1. First campaign: The Cohetero stream and the Juan N. Álvarez Avenue cleaning action.

   The Cohetero stream is the main conduit for sewage contamination in the municipality of Atoyac. In this campaign, students from different primary schools participated. The citizen and press pressure due to the first campaign rose the government action, as we could see in the following operations.

2. The Paraiso beach lagoon cleaning campaign.

   Approximately 800 citizens participated, among them the mayor of Atoyac de Álvarez and the municipal president of Benito Juárez.

3. The Paraiso River section cleaning action.

   The presidents of Atoyac de Álvarez and Benito Juárez participated.

4. The Teotepec Hill reforestation.

   Approximately 3,000 participated – among them the presidents of Atoyac de Alvarez, Benito Juárez and San Miguel Totoloapan.

The megaproject "Las Regiones Sociales en el Siglo XXI", also called Pro-Regiones, was launched in July 2005 as part of the "Sociedad y Cultura: México Siglo XXI" program, promoted by UNAM. The main objective of this megaproject is to use the potential of the social sciences to raise the standard of living of Mexicans. The members of the Citizen Council for the Atoyac River Rescue collaborated with the Pro-Regiones coordinators, who could carry out numerous studies on solid waste and sewage management.

In the following years, acts of violence and organized crime increased in the country, making it difficult for the scientists and ecologic activists to continue their field work. This scenario discouraged the citizenry, who now worried about staying safe from organized crime.

It was not until 2015 that the collective "Guerrero es primero" was created as an initiative of different social organizations to face of the difficult panorama of Mexico and the situation of violence in the state of Guerrero. The objective of this collective is to promote spaces for dialogue and the search for alternatives and collaboration in order to contribute to the construction of peace, the full exercise of human rights and sustainable development in the region. Their actions are based on communication and the construction of consensus and collaborative work agendas between actors from different sectors.

Figure 6.8 The Paraiso lagoon cleanup campaign was the second action undertaken
       by citizens

In his article "Guerrero: The Circular Violence", Carlos Illades (2015)
provides the following description.

> Since its formation in 1849, the state of Guerrero has been poor, iso-
> lated, quite socially unequal, cacique and with precarious political sta-
> bility. Also, rich in natural resources and biodiversity, multicultural,
> rough and with a strong tradition of popular organization.
>   The same author also points out that the state shares with Oaxaca
> and Chiapas the lowest social indices in the country, but, unlike them,
> its strategic position in the transfer of drugs, in addition to being one of
> the main world producers of gum poppy, they have placed it in a quali-
> tatively worse situation than those entities, particularly with regard to
> criminal violence.
>
>   (Illades, 2015)

Actions in favor of the Atoyac River were reactivated at the beginning of
2017, when a group of citizens – which would later become the "Salvemos
el rio Atoyac" collective – carried out environmental education workshops
and campaigns to clean up the river in the urban area. This last act had positive repercussions in the political and social spheres, since it was a wake-up
call about the ecological situation of the basin that led to a consultation in

the municipal council whereby citizens; various actors from the agriculture, livestock and forestry sectors; and municipal authorities participated.

Sanitation Recommendation du to Human Rights Violation

> Water is a basic human right and is fundamental to human dignity . . .
> Today, three out of ten people in the world do not have access to safe
> drinking water. Six out of ten lack safely managed sanitation services . . . .
> If people cannot enjoy their right to water, they cannot enjoy their right
> to life.
>
> Michelle Bachelet, United Nations High Commissioner
> for Human Rights, March 19, 2019, Declaration on
> World Water Day: "Leave No One Behind"

According to the United Nations, economic, social and cultural rights include the rights to food, adequate housing, education, health, social security, participation in cultural life, water, sanitation and work.

Under this premise, members of the citizens organization "Guerrero es Primero" met with the National Commission of Human Rights (CNDH) authorities in 2016 to express their disagreement on the Atoyac River pollution and its consequences on the environment and the health of the population.

In August 2019, the CNDH issued a recommendation addressed to the authorities of the municipalities of Atoyac de Alvarez and Benito Juarez. In this long document, they pointed out the authorities responsible of the degradation of the ecological state of the river and established the steps and time to solve some of the main environmental issues of the basin.

The Collective "Salvemos El Rio Atoyac"

This collective was created by the citizens of Atoyac de Alvarez in 2017 to unite efforts for the restoration of the river. Its main objective was to form a community of citizens of all ages and professional horizons eager to be part of the change, with the aim of converging initiatives and actions for the ecological improvement of the Atoyac River and surroundings.

Its proposed lines of action were as follow.

Communication and Public Awareness

These efforts are to achieved by developing campaigns of communication via social media, radio and local television broadcasts, and by manifesting in public places; production of short films all along the river basin to record the ecological and social problems; and preparation of documents based on scientific data directed to authorities and experts on the subject

Figure 6.9 Citizens of all ages gathered April 21, 2021 in the municipal plaza to defend environmental rights and denounce the contamination of the Atoyac River and the inaction of the authorities

Environmental Education

The objective of the environmental education programs is to educate and raise awareness among youth and adults by the following methods.

- The discovery of biodiversity and the natural heritage of the region.
- Understanding the ecological interactions of the species in riverine and coastal ecosystems and the impact of human activities on these interactions.
- The development of a sense of belonging and responsibility in caring for the surrounding environment

Cleaning and Reforestation Campaigns

This involves organization of cleaning and reforestation campaigns in the most damaged zones of the river with the participation of students, citizens and local authorities.

Figure 6.10 Environmental education

Figure 6.11 The technique of "photolenguage" during a workshop with high school students

Figure 6.12 Creation of banners about the preservation of the river after one of the workshops in a primary school

Citizen Science Program: Co-participatory Strategy for Water Management in the Atoyac River Basin, Guerrero

Madrid E., Sampedro L., *Maganda C.

Centro de Ciencias de Desarrollo Regional – Universidad Autónoma de Guerrero, CCDR-UAGRO.

*Instituto de Ecología, A.C- INECOL

esthermorales@uagro.mx, 05156@uagro.mx, carmen.maganda@inecol.mx

Figure 6.13 First cleaning campaign of the collective in 2017

Figure 6.14 Cleaning and reforestation campaign of the collective in 2021

This research addresses the problematic contamination of the Atoyac River Basin in Guerrero and the way in which river neighbors/citizens relate to the quality of their water. The objective is to link a water management project in a participatory way with inhabitants of the communities near the main river (from the upper, middle and lower parts of the basin), and with various actors from the government sector, civil organizations and academia. The project includes a diagnosis phase with participative workshops to identify the main socio-environmental problems and determinate the quality of surface water through the use of physicochemical parameters and bioindicators. A second phase will develop the co-generated and participatory action strategy with members of the selected riverside communities.

The methodological approach is mixed and is carried out through an interdisciplinary approach using participatory methodologies to promote social bonding and experimental activities at the laboratory level for the determination of water quality with physicochemical parameters and bioindicators.

Figure 6.1 shows the geographical location of the Atoyac River Basin and the nine points selected for the sampling of physicochemical parameters and bioindicators. The sampling is carried out during a year to include the rainy and dry seasons. The physicochemical parameters to be determined based on the Traffic Light of the Mexican Water Commission (the methodology they use to measure the water quality) include: temperature, pH, total dissolved solids, electrical conductivity, nitrates, nitrites, ammonia nitrogen, sulfates, dissolved oxygen and biochemical demand on the fifth day. Specific sampling is also included to determine the presence of heavy metals at the selected points. The determination of the water quality, through bioindicators, is carried out by

Figure 6.15 Citizen science program: Co-participatory strategy for water management in the Atoyac River basin

observing the physical characteristics of the shoreline and collecting aquatic macroinvertebrates at four times of the year, which include the dry and rainy seasons, in the nine selected points. Both procedures have always been accompanied by local actors. We selected four of the nine localities to share the results of the obtained progress to provide training through participatory workshops with pro-environmental issues and to promote collaboration towards the co-participatory strategy of water management. The four selected localities correspond to the upper, middle and lower parts of the basin.

Monitoring of Policies Related to the Restoration of the River

The members of the collective act as the social counterpart to give timely follow up to the CNDH Recommendation No. 56/201, demanding compliance and transparency of the suggested actions and permanently informing the citizens about the progress.

Other activities include the sampling and monitoring of the river water quality all along the basin.

Learning and Perspectives

The guarantee of the right to a healthy environment has evolved from international human rights law to national constitutions; however, given the insufficiency of this protection, collective actions have been developed aimed at protecting collective and diffuse interests and achieving the repair of environmental damage.

Environmental protection is also a vital part of contemporary human rights doctrine, as it is a *sine qua non* for numerous human rights, such as the right to health and the right to life. There is hardly any need to elaborate on this, since damage to the environment can undermine and weaken all human rights in the Universal Declaration and other human rights instruments.

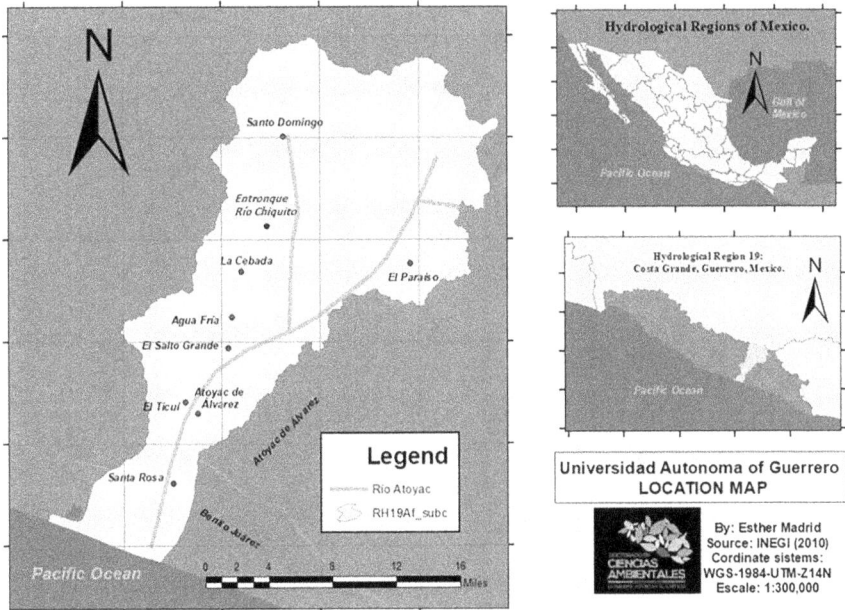

Figure 6.16 Location map of the study area in the Atoyac River Basin, Guerrero

Figure 6.17

Figure 6.18 The monitoring of the river water quality all along the basin

In the different regions of the state of Guerrero, where the corruption of government institutions and indifference to environmental problems have been present for decades, citizen action plays the role of bearer of alert and catalyst for changes in the relationship between society and the environment.

The relatively recent consolidation of the collective "Save the Atoyac River" stems from the need to take into hands the issue of environmental injustice caused by the contamination of the Atoyac River. The volunteers come from varied backgrounds but all work for a single objective – the rescue of the Atoyac River, and with it the restoration of its biological, ecological and social roles.

If we look at the problem of contamination of the Atoyac River and the role that each of the social actors has played, we realize that the citizen initiatives are the ones who have managed – despite the errors and ups and downs – to generate proposals and actions for protection of the environment.

According to the experience of the work carried out to date, in order to implement and manage an environmental project, the following points need to be included.

1. Investigate the problem from different angles – ecological, social and economic – taking into account the point of view of the different socio-economic actors.
2. Build structured and scientifically supported proposals that give rise to concrete actions that involve the spheres of society.
3. Educate and raise awareness among the population in general, adapting the discourse, strategies and tools to each age and profile.
4. Create work networks in the different communities and motivate the participants with campaigns and field actions that involve them and create a feeling of appropriation and belonging to the territory.

Bibliography

Anglés Hernández, Marisol (2015) Acciones colectivas en materia de protección ambiental fallas de origen, *Boletin Mexicano de Derecho Comparado*, 48(144), 899–929.

Bachelet, Michelle (2019) The United Nations world water development report: Leaving no one behind. UNESCO World Water Assessment Programme.

CNDH, Comisión Nacional de Derechos Humanos (2019) Recomendación no. 56/2019, Sobre el caso de las violaciones a los derechos humanos a un medio ambiente sano y al saneamiento del agua, por la contaminación del Rio Atoyac.

CONAGUA, Comision Nacional del Agua (2012) Situacion del subsector agua potable, alcantarillado y saneamiento. SEMARNAT, Mexico, D.F., p. 91.

Fernandez Gomez, R. y Fierro Leyva, M. (2015) Participacion ciudadana y desarrollo sustentable en la rcuperacion de la Cuenca del Rio Atoyac: Proyecto Pro-Regiones UNAM-UAGro. *Tlamati*, 6(4), 59–62.

Fierro Leyva, Martin (2015) Mecanismos de participación para el saneamiento ambiental de la cuenca del río Atoyac, Municipios: Atoyac de Álvarez y Benito Juárez. *Estado de Guerrero, México, REVETEDE Revista Electronica de Educacion, Tecnologia y Derarrollo Regional*, 1(1).

Illades, Carlos (2015) Guerrero: La violencia circular. *Nexos*, 2014, 1–15.

Jiménez Varela, A., Méndez Bahena A., Alvarado Gomez A.V. y Rivera Barreto C. (2000) Diagnostico amiental de la cuenca del Rio Atoyac, Guerrero. *Altamirano*, junio–julio, 16, 43–65. Disponible en: Revista%20Altamirano%20No.16.Jun-Jul%202000.pdf. Ultima consulta: Diciembre 20, 2018.

Pro-Regiones (2007) Hacia el rescate de la cuenca del rio Atoyac. Proregiones UNAM-IIS-UAG-IIEPA-IMA, Mexico.

SAGARPA Secretaria de Agricultura, Ganaderia, Desarrollo Rural, Pesca y Alimentacion (2016) SIAP Servicio de Informacion Agroalimentaira y Pesquera. Estadisticas de la produccion agricolaa de 2016. SAGARPA, México. Disponible en http://infosiap.siap.gob.mx/datosAbiertos.php. Ultima consulta: Septiembre 24, 2018.

Webography

http://ri.uagro.mx/bitstream/handle/uagro/550/11995_ART2015OK.pdf?sequence=1&isAllowed=y

https://www.ohchr.org/SP/Issues/ESCR/Pages/Water.aspx

http://www.conabio.gob.mx/conocimiento/regionalizacion/doctos/rhp_028.html

http://cuentame.inegi.org.mx/territorio/agua/sobreexplota.aspx?tema=T

https://files.conagua.gob.mx/conagua/generico/calidad_del_agua/diagnostico_atoyac_guerrero_2012-2019.pdf

7 "The Kids Will Be Alright" – Interweaving Urban Design, Youth Perspectives, and Sustainability Education

A Case Study of the New International School of Japan

Sebastian Krutkowski and Lionel Rogers

Introduction

Urban design concepts are typically integrated into secondary school curricula in courses such as geography, social studies, and environmental science. The relationship between urban design and climate change is increasingly recognised, with a focus on how city planning can either contribute to climate change mitigation or exacerbate its impacts. In environmental science courses, discussions often delves into urban ecology and sustainable development, exploring strategies for climate resilience such as green infrastructure and energy-efficient buildings. Some schools may adopt interdisciplinary approaches, connecting urban design concepts with subjects like mathematics, architecture, or economics. Elective courses may offer more in-depth explorations of urban design, encouraging students to engage with real-world challenges and propose solutions, fostering awareness and community engagement. We advocate that due to the dynamic nature of the field, coupled with the global emphasis on sustainability, secondary school curricula should evolve to address the critical role of urban design in climate change adaptation and mitigation, in order to prepare students for a future when sustainable urban development is a crucial aspect of addressing environmental challenges.

We posit that the "urban" is the uttermost example of human influence on our physical environment. Often characterised by densely populated communities and intensive infrastructure, urban areas around the world significantly alter natural ecosystems, contributing to alterations in land use, biodiversity loss, and increased carbon emissions (McGranahan et al., 2005; Alberti, 1999). The ever-changing and intricate world of city living necessitates careful consideration when it comes to dealing with climate change. Realising how important cities are in the broader environmental context highlights the importance of designing and planning cities in a sustainable way to tackle the

DOI: 10.4324/9781003486961-7

issues associated with urban growth (Romero-Lankao et al., 2012; Pickett et al., 2001; Kennedy et al., 2016).

Within the broader discourse of urban influence on the environment, the concept of urbanisation extends beyond population density and infrastructure. Urban sprawl, i.e. the uncontrolled expansion of cities into surrounding areas, poses challenges to sustainable development (Burchell et al., 2002). Simultaneously, gentrification – often associated with revitalisation efforts – can lead to the displacement of marginalised communities, raising concerns about social equity (Lees, 2012). These urban dynamics intertwine with issues of diversity and inclusion, as marginalised groups face disproportionate impacts due to environmental changes and planning decisions (Wolch et al., 2014). Recognising these interconnected challenges is pivotal for promoting urban resilience and fostering inclusive, equitable urban environments. Therefore, as cities evolve and grow, addressing the complexities of urban sprawl, gentrification, and issues of equity becomes imperative for sustainable urban development (Grant & Perchoux, 2019).

Well-designed cities with green infrastructure enhance climate resilience, mitigating extreme weather impacts such as hurricanes, floods, mudslides, cold snaps, or heat waves (Beatley, 2011). Compact, multi-purpose designs cut transportation needs and promote energy efficiency, reducing carbon emissions (Kennedy et al., 2016). Sustainable urban practices also include energy-efficient buildings, renewable energy, self-sufficiency and effective waste management. Furthermore, strategic land-use planning safeguards biodiversity and native wildlife, and preserves carbon-absorbing ecosystems. Finally, liveable and walkable communities with accessible public transport reduce individual car use, easing emissions and traffic congestion while also boosting individual health of urban dwellers. One of us had the opportunity and pleasure to supervise a senior student's capstone project which is a relevant case in point here. Their project delved into global infrastructure issues, focusing on Singapore. The student took the initiative to contact the Singaporean government, specifically inquiring about the nation's strategies for traffic prevention. To our surprise, the inquiry garnered a response from Siti Rohani Kasmany, the Assistant Manager of Quality Service Operation for Land Transport Authority. Kasmany highlighted the evolution of Singapore's approach to managing vehicular ownership and usage since the 1970s:

Since the 1970s, Singapore has progressively implemented a variety of vehicle ownership and usage restraint measures. Vehicle ownership restraints are mainly in the form of fiscal disincentives such as import/excise duties (ID), Additional Registration Fees (ARF), and road taxes. Usage restraints include the Area Licensing Scheme, (ALS) and Electronic Road Pricing (ERP).

Regarding the ALS, introduced in 1975, Kasmany noted:

> Before the introduction of ERP for congestion pricing and the Certificate of Entitlement (COE) scheme for vehicle population control, Singapore implemented as the predecessor of ERP, the ALS. It was first introduced in 1975 for road pricing within the Central Business District. The ALS was a paper licence scheme, which required motorists to pre-purchase a paper licence to enter the Restricted Zone (RZ). Human sentries at the ALS gantries would enforce the scheme by manually checking the windscreen of vehicles entering the RZ for a valid licence.

Explaining the ERP system, they further remarked:

> ERP rates are reviewed regularly and are adjusted according to prevailing traffic conditions. ERP rates increase when traffic speeds are below optimal levels, while ERP rates are reduced when traffic speeds are above optimal levels. This mechanism seeks to keep traffic speeds at optimal levels vis-à-vis the number of cars on the roads. The ERP system remains an effective tool where data indicate that speeds have generally improved when ERP rates are increased, and vice versa.

In light of these insights, it becomes evident that incorporating environmental considerations into urban planning is essential for effectively addressing local climate change challenges.

The student's initiative to explore and engage with global urban infrastructure challenges, exemplified through the investigation into Singapore's traffic management strategies, demonstrates exceptional curiosity and proactive learning. Their dedication to reaching out to experts in the field, as evidenced by the response from Siti Rohani Kasmany, showcases a commendable commitment to understanding and addressing real-world urban issues among today's youth. This experience underscores the value of students proactively engaging with real-world urban challenges. It highlights the practical application of theoretical knowledge and the potential for valuable connections with industry experts. Initiating dialogues with professionals enriches the educational experience, providing students with practical insights and fostering a deeper understanding of complex urban issues.

In the face of the complexities of urban challenges today and tomorrow, the role of young people in urban design emerges as a catalyst for transformative solutions. Despite urban dynamics like sprawl and gentrification, youth – often relegated to the sidelines as "future citizens" – possess untapped potential for shaping inclusive urban futures. After all, building cities of the future should be built with future citizens in mind. Integrating young perspectives into the planning process becomes an avenue for infusing creativity, energy, proactive concerns, and fresh ideas, countering conventional notions of urban development. As cities grapple with the intricacies of sprawl and gentrification, harnessing the contributions of young individuals becomes not only an opportunity for innovation but a necessity for building truly vibrant, equitable, and sustainable urban spaces. Some of the ways our youth can contribute to a sustainable urban future are through engaging in public art and expression, advocacy, community engagement, temporary and "tactical urbanism", collaboration with local authorities, or low-scale green and sustainable initiatives on school campuses.

One thing we observed in our time as humanities and social studies teachers is that young people view urban design as a critical element in addressing climate change, both in terms of adaptation and mitigation. They recognise the urgent need for cities to transform their infrastructure to become more resilient to the impacts of climate change. From our experience, young individuals often emphasise the importance of sustainable and eco-friendly urban planning, promoting the use of green spaces, renewable energy, and low-carbon transportation options. This is why young people around the world may increasingly wish to advocate for inclusive and community-centric urban design that ensures social equity in the face of climate challenges. By actively participating in discussions, appealing for sustainable policies, and contributing innovative ideas, young voices are shaping the dialogue around urban design, emphasising its pivotal role in creating climate-resilient cities for the future.

In the context of our school, located in the heart of Tokyo – one of the world's largest metropolitan hubs – the young people we listen to in class specifically call for an age-friendly urban environment, one that is tailored to the needs of 8-year-olds as well as 80-year-olds. According to Danish architect Jan Gehl (2010), including the young and the old is the ultimate seal of approval in urbanism – seeing many children and many old people using the public spaces in the city is a sign that there's a good quality of life for people there. Creating spaces that cater to the diverse needs of different age groups will ensure accessibility, safety, and amenities that enhance the quality of life for both children and seniors. By focusing on this dual-age approach, urban planning can create spaces that are not only inclusive but also address the unique requirements of both the younger and older populations, promoting a more harmonious and liveable city for everyone.

In this chapter, we want to amplify the voice of the 12–18-year-olds, many of whom are "third-culture kids" (TCKs), i.e. individuals who, during their formative years, have lived outside their parents' home culture due to factors like parental work assignments (Useem et al., 1963). This experience leads TCKs to blend aspects of their home culture, their current residence culture, and a unique amalgamated third culture developed from their diverse cultural exposures. TCKs often develop a distinctive identity that sets them apart from their parents' and host cultures, creating a "third culture" shaped by their multi-cultural upbringing.

The Context of Our School

Since its establishment in 2001, New International School of Japan (NewIS) has been at the cutting edge of challenging many structural preconceptions of schooling, such as age-grade classes, separation of languages, language usage rules, and the role of summative assessment. Constructivist philosophies, such as those of Vygotsky (1965) and Engeström (1987), drive and shape our learning approach, centring around the idea that language precedes thought and holds the power to enrich one's emotional understanding and experience. NewIS, by design, is a multi-age, Japanese and English dual-language learning community located in the Tokyo metropolitan area, providing education for 280 3–18-year-olds.

NewIS uses an adapted, re-contextualised version of Scotland's Curriculum for Excellence (CfE) – a holistic and innovative educational framework designed to empower learners to become successful individuals, responsible citizens, and effective contributors in a rapidly changing world. Enacted in 2004, the CfE marks a departure from traditional education models, emphasising a learner-centric approach that seeks to cultivate skills and values essential for life, work, and global citizenship.

The CfE social studies experiences and outcomes are structured into three main organisers: people, past events, and societies; people, place, and environment; and people in society, economy, and business. These main organisers acknowledge the unique contributions of each social subject while accommodating local contexts. Teachers can use this framework to foster interdisciplinary learning, making connections between subjects. The main organisers provide flexibility for teachers to explore opportunities for collaborative planning and enhance learning within and across curriculum areas, supporting the development of coherent programmes of learning within and between educational institutions.

In the people, place, and environment organiser, several key points relevant to urban design and sustainability education emerge:

Interconnectedness of people and place, including cultural and social understanding:

- Emphasises the dynamic relationship between individuals and their surroundings.
- Encourages learners to explore how people shape and are shaped by their environment.
- Focuses on developing an appreciation for cultural diversity and social structures within different environments.
- Encourages learners to recognise the influence of history and traditions on contemporary societies.

Global citizenship and sustainable lifestyle choices:

- Promotes the concept of global citizenship by fostering an understanding of the interdependence of people and places worldwide.
- Encourages students to consider their roles and responsibilities in a global context, with an emphasis on sustainable choices.
- Encourages an exploration of environmental issues and solutions, aligning with broader sustainability education goals.

Experiential and outdoor learning:

- Supports hands-on, experiential learning to deepen understanding.
- Encourages outdoor learning experiences to connect theory with the physical environment.

These key points reflect the comprehensive goals and principles embedded in the "People, Place, and Environment" organiser, contributing to the development of informed, responsible, and globally aware citizens prepared for active participation in society. Using this as the curricular basis for our classes, our overarching learning intentions were to develop students' analytical skills to understand the complex relationships between people, place, and environment. We strived to promote a "place-based" approach, connecting learning experiences to the geographical and cultural context of the students, fostering a commitment to lifelong learning in the process. Having taught urban design to 140+ students over the past two years, we hope to share our approach and argue how it can help advocate for the incorporation of local and global examples in the learning process. The starting point is to recognise the importance of interdisciplinary skills in addressing real-world challenges.

We are obliged to note one limitation to our case study – namely, that discussions on urban design and mental health may differ between English and Japanese contexts due to cultural, linguistic, conceptual, and societal subtleties. Cultural perceptions of mental health, urban design philosophy, and community emphasis may shape the discourse in unique ways. The linguistic nuances between English and Japanese could influence the expression of mental health concepts (or lack thereof) and urban design principles. The differences can also be influenced by the history and characteristics of different cities, the rules and plans governments have, and how the public gets involved. In Japan, the tendency of individuals to adjust their behaviour, attitudes, or beliefs to align with the prevailing social norms and cultural values such as harmony may lead people to suppress their individual needs or preferences; as a common cultural adage goes, "出る釘は打たれる" or "The nail that sticks out gets hammered in". This adherence to conformity may influence urban design priorities for mental well-being. Recognising these differences is crucial for fostering a culturally sensitive dialogue that addresses mental health considerations within the specific context of each language and culture. Balancing these nuances will be essential in our future classes for effective urban design interventions that promote mental health and well-being in both English-speaking and Japanese communities.

What follows is the lessons learned from urban design classes we delivered at NewIS over the course of the last two academic years.

Psychology of Urban Design (Spring Term 2020–2021)

The main challenge, which links to our rationale for developing urban design learning experiences at our school, is the population structure of the city at large. As of October 1, 2015, Tokyo's population stood at 13.515 million, with 11.5% being children (ages 0–14), 65.9% in the working-age bracket (ages 15–64), and 22.7% categorised as aged, i.e. 65 and over (Tokyo Metropolitan Government, 2023). Notably, Tokyo surpassed the United Nations' 14% threshold for an "aged society" in 1998, with seniors now constituting 21% or more of the city's population, marking Tokyo as a "super-aged society". An ageing population in Ikebukuro (and Tokyo as a whole) requires a targeted urban strategy to address the unique needs of different demographic groups, ensuring inclusivity and sustainable development in the metropolis.

Over 56% of the world's population now resides in cities (World Bank, 2023), underscoring the importance of examining not only the physical aspects of urban design, such as architecture and infrastructure, but also the psychological impact that urban living has on people. NewIS is located in the Ikebukuro area of Tokyo, one of the city's main transit centres. Characterised by a high-density built environment, the area around the school is also a bustling commercial and entertainment hub. Ikebukuro boasts a diverse population structure with a mix of residents, office workers, and visitors, contributing to its dynamic and cosmopolitan atmosphere. However, the area also faces urban design challenges, including congestion, limited green spaces, and occasional overcrowding in transportation hubs, affecting overall liveability. The diverse architectural styles, while adding character, contribute to a somewhat fragmented urban aesthetic affecting people's social and emotional well-being. High-density and commercial activities elevate noise and air pollution, posing further challenges. Ageing infrastructure and safety concerns compound these issues. Addressing these challenges requires strategic urban planning to balance the dynamic environment and enhance Ikebukuro's overall quality of life.

Why Psychology?

A city, much like a person, is composed of different parts and develops its qualities over time. Each city possesses a distinct soul and unique personality, capable of flourishing or deteriorating based on the urban design choices we make. Bad planning and architecture can induce anxiety, alienation, or depression among city dwellers and even influence laws. Conversely, good urban design, informed by psychology and the study of our emotional behaviour, can enhance the liveability of our cities. This is why we created the class, recognising our students' need for heightened awareness regarding their interactions and behaviours within their immediate environment. Our decision stemmed from the realisation that certain behaviours were influenced and predetermined by the design elements surrounding them. Hence, we commenced with a focus on the intricate relationship between human behaviour and the built environment through the lens of "the psychology of urban design". As part of the class, we investigated the unique personalities inherent in cities through the administration of a distinct City Personality Test (Urban Psyche, 2020), which allowed our students to scrutinise the criteria contributing to the designation of the most liveable cities, affording a comprehensive framework to evaluate and comprehend the diverse factors shaping the quality of urban life.

At the start of our teaching journey, although we wanted to dive into the particulars of urban design, we decided to start with the simplest thing we could imagine – a casual stroll around our school neighbourhood. We actually did this several times, the first time to gauge their initial reaction and understanding of the world and structures around them, and several other times as we introduced new concepts to see how much change in their awareness occurred from the initial walk. In the beginning, the students' responses were quite basic reactions.

> I felt good walking outside and watching the city structure. It has helped me to understand what kind of design our city has. I can apply these skills to make my decision where to live one day.

Yet as new ideas and world building elements were introduced to them, we began to see the lessons take shape.

> From my personal opinion, actually getting out of the class and going to places to observe was very interesting. Some people was just having fun of the concept of getting out of the class and fun talking with their friends, but I was personally surprised of the streets of Ikebukuro, because I've been in this school for more than ten years and been walking the streets more than anyone and I haven't realised how it changed, and how fascinating it is to understand the current Ikebukuro (still dirty and smelly). Also, I was personally happy to expand the knowledge of how nature is an important concept to human society.

One piece of vocabulary that really stuck out to the students in the course of the term was the acronym GAPS (Green, Active, Pro-Social, and Safe space) and the attributes of each initial. This one acronym, it seems, finally gave the cohort members the vocabulary they lacked for concepts they innately knew. We taught them that to identify spaces that would promote well-being mentally all they had to do was "fill in the gaps". Many students referred to this in their reflections, e.g.:

> I think taking a walk and looking around Ikebukuro helped me pay attention more to how humans interact with buildings and if the GAPS are filled.
>
> Most people need green or blue places, active places, prosocial places, and safe places around them. The effect of these places is to open our hearts, become healthy, calm down, etc.

This perspective underscores the significance of creating environments that fulfil these fundamental human needs, contributing to overall well-being and emotional states. As students agreed that nature or exposure in some form boosted their emotions and mental state, it only seemed natural that we also included the theories of biophilia and attention restoration (Kaplan, 1995; Berman et al., 2008).

The activities centred around the GAPS philosophy confirmed how it is an imperative to consider greenery, safety, and opportunities for physical exercise and social interaction to foster vibrant and sustainable urban environments. Well-designed urban spaces necessitate the integration of these key elements to cater to the diverse needs of residents and visitors. The provision of green spaces not only enhances aesthetics but also contributes to environmental sustainability and well-being (Ward Thompson et al., 2012). Active spaces, conducive to physical exercise, promote a healthier lifestyle and contribute to the overall fitness of the community (Sallis et al., 2012). Moreover, spaces that encourage pro-social interactions are essential for building a sense of community and connectedness (Gehl, 2010). Safety measures are equally crucial, ensuring that these spaces are welcoming and secure for all users (Coley et al., 1997).

As this was our first class in a series of instalments centred on urban design, we made sure we introduced our students to pressing issues in urbanism such as sprawl and gentrification, alongside contemporary concepts like the innovative 15-minute city model proposed by Carlos Moreno (O'Sullivan & Bliss, 2020). Grounding our study in empirical observation, we systematically examined the immediate school environs, seeking to discern the palpable impact of design choices on the human experience. The nuanced understanding of hostile architecture and the psychological ramifications of varying scales in urban planning became integral facets of our inquiry. We facilitated a structured engagement with creative photo assignments, allowing students to visually document and analyse the symbiotic relationship between design and human behaviour. Following are some images of the "human scale" photo assignment we have set in the early weeks of the term.

This was not just about throwing facts at our students; it was about giving them a real sense of the challenges and possibilities that come with shaping urban spaces. We wanted them to see the bigger picture. So, we delved into these issues, not just as topics in a syllabus, but as real challenges that cities grapple with. This laid the groundwork, setting the stage for more in-depth discussions on the psychology behind urban design.

Reflecting on this, one student shared,

> I was able to analyse issues surrounding urban design such as gentrification, homelessness, as well as policies of containment, and escaping from the problem by not dealing with the root cause.

Figure 7.1 Human scale homework, example 1

Figure 7.2 Human scale homework, example 2

Figure 7.3 Human scale homework, example 3

Figure 7.4 Human scale homework, example 4

This insightful observation resonates with our pedagogical approach – encouraging students to critically examine and comprehend the intricate layers of sustainable urban design, moving beyond the surface to address the root causes of urban challenges.

Another student added,

> By furthering my understanding of a variety of urban design practices and principles, I was able to independently observe, analyse, and evaluate how theories such as GAPS, the 15-minute city, and urban sprawl were affecting my community and mental health.

This emphasises the empowerment we aim to instil in our students, enabling them to apply theoretical knowledge to real-world scenarios and fostering a proactive engagement with urban design principles for the betterment of their communities.

There is a degree of unpredictability in classroom learning. Our discussions of the GAPS philosophy took us into considerations of *lighting* as a factor influencing mental health and behaviour, and *light* as a "material" in architecture (Libeskind, cited in Alda, 2023). This underscores the dynamic nature of urban design education, whereby unexpected insights can emerge, enriching both teachers and students in understanding the psychological dimensions of built environments. There is a reciprocal learning process between teachers and students, leading to a broader understanding of the psychology of urban design. A poorly designed classroom with inadequate light, which was an "icebreaker" discussion prompt in one of our lessons, served as a tangible entry point for students to grasp the broader concept of how the physical environment affects their experiences. One of our students developed a significant interest in the role of light as a crucial building material in architecture, as well as a factor with profound impact on mood and well-being. He wrote his senior project essay that year on how light, both natural and artificial, plays a transformative role in shaping and defining spaces. Exploring the psychological use and impact of light, the student discussed how light influences mood, productivity, and well-being. Additionally, the author advocated for architects and designers to prioritise a holistic approach to lighting, recognising it as a fundamental material that significantly contributes to the success of a built environment.

As mentioned before, we aimed in this class to ensure that students grasped the dual role of urban design – its psychological impact on daily life and the profound influence on mental health. Notably, throughout the term, this revelation was as eye-opening for us as teachers as it was for the students. So much so that we replicated this section with our junior high crowd a few years later, a point we will delve into later in this paper. Initially, the students did not connect changes in mood with alterations in their environment. To illustrate, we used a straightforward example – examining the correlation between the classes they took and their locations. Ironically, the room we used for this class was the least favoured among both students and teachers. Its ill-shaped floor plan, cramped quarters, lack of natural light, and inability to sustain plant life, coupled with the absence of windows and views, made it the top contender for the worst room in the school. During a follow-up discussion about how students choose their classes, we inquired about the factors they considered. The initial responses included the topic of the class and, unsurprisingly, the instructor. When asked about the location of the class, initially, no correlations were made by the students regarding its impact. It was not until someone expressed, paraphrased, "some places just feel better, and that [other] class, in particular, feels like a prison", that a sliver of realisation sparked broader awareness.

Art and Culture as Revitalisation

The subsequent case studies we referred to included examples of shutter art from cities around the world, including London's "Paint the Town" (London First, 2017) and "Legal Shutter Tokyo" initiatives. To our surprise, two of our students revealed themselves as the authors of a prominent shutter artwork in the very heart of the city – the famous Shibuya Scramble. Shutters decorated with creative designs can make an area more attractive, especially in the evening. Students believe that injecting bright colours into a previously

Figure 7.5 Art rejuvenation example – Tsutaya Shibuya shutter art

Figure 7.6 Art rejuvenation example – Kawasaki Festival 1

Figure 7.7 Art rejuvenation example – Kawasaki Festival 2

Figure 7.8 Art rejuvenation example – Kawasaki Festival 3

Figure 7.9 Art rejuvenation example – Kawasaki Festival 4

grey streetscape can considerably improve safety and the community's sense of pride. The following mural was heavily influenced by the pandemic era and the numerous restrictions government authorities in Japan imposed – both locally and nationally. The authors of the artwork continued to organise workshops in youth centres (e.g. Daikanyama Teens Creative) on occasion of cultural festivals and events.

Our students would often mention how the COVID-19 pandemic put a strain on their mental well-being, mainly due to the inability to meet their friends and move freely around Tokyo and beyond. This is when the importance of simple, grassroots initiatives in public spaces became more evident when we studied cases of urban renewal following natural disasters. For instance, students enthusiastically responded to the examples of urban renewal in post-earthquake Christchurch, New Zealand – all part of the portfolio of the Gap Filler community service organisation. In the aftermath of the 2010 Ōtautahi earthquake, Gap Filler launched a series of successful public events and activities to support people in crisis and help create a new identity for the city. Gap Filler has since grown from simple beginnings via actions that exemplified guerrilla-style adaptive urbanism into an international creative placemaking consultancy. The organisation's urban regeneration projects in Christchurch represent a dynamic interplay between creative placemaking and urban design principles. The initiatives showcased on their portfolio, such as the Pallet Pavilion and Dance-O-Mat, transcend traditional urban spaces, embodying the GAPS philosophy. The Pallet Pavilion, constructed from reused materials, not only aligns with sustainability principles (Gehl, 2010) but also serves as a vibrant, pro-social gathering space, fostering community interaction (Coley et al., 1997). Likewise, the Dance-O-Mat transforms a vacant site into an active space, encouraging physical exercise and social engagement (Sallis et al., 2012).

Gap Filler's projects exemplify the ethos of inclusive urban design by repurposing underutilised spaces. The Book Exchange, a miniature outdoor library, exemplifies the integration of greenery and social interaction, promoting a sense of community (Ward Thompson et al., 2012). By activating vacant lots, Gap Filler contributes to urban resilience, embracing Gehl's (2010) vision of cities for people. These projects serve as practical examples of how temporary interventions can have lasting impacts, emphasising the importance of adaptive, community-centric urban design in enhancing the liveability and vibrancy of urban spaces.

Another example we discussed in class was the "I Wish This Was" project, initiated by Candy Chang in 2010. The artist blends street art and urban planning by placing stickers on vacant buildings in New Orleans (after Hurricane Katrina), prompting residents to express their desires for these spaces. Chang's participatory experiment encourages on-site civic engagement, yielding diverse responses from practical to poetic, such as wishing for a butcher shop, community garden, or a place for conversation. Presented at the 2012 Venice Architecture Biennale, the project has transcended New Orleans, inspiring global communities to envision and shape the potential transformations

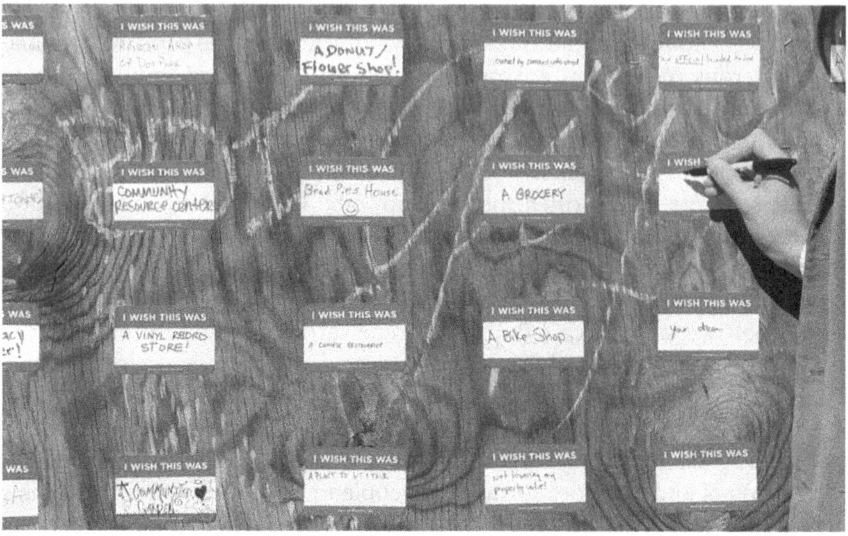

Figure 7.10 "I Wish This Was" Project in New Orleans
Source: Image courtesy of https://candychang.com/work/i-wish-this-was/

of their own spaces, offering a unique perspective on the future of communal development. "I Wish This Was" and Gap Filler's Christchurch projects both engage communities in transforming urban spaces. "I Wish This Was" uses stickers on vacant buildings for residents to express visions, while Gap Filler regenerates post-earthquake sites with creative installations. Both initiatives highlight the necessity of community involvement in revitalising and shaping urban environments.

While our school does not have a campus *per se*, the space between buildings that belong to our campus is lively and often shared with local residents. These spaces regularly "come alive" during school events such as the Spring Festival or special seasonal celebrations like *mochitsuki* – a community gathering to make *mochi*, a kind of rice cake that symbolises good fortune. Even when there is no direct reference to a calendar event, initiatives within school premises benefit the local community, e.g. greenery, the herb garden, and ponds maintained by students in the "Tokyo Wildlife" class.

City Personalities

The "Psychology of Urban Design" class ended with a commitment to schedule a follow-up, which we will go through in the next section of this chapter. The final activity in the class featured students taking the "City Personality Test" (Urban Psyche, 2020), a tool which helps to imagine what would happen if a city could take a personality test to determine traits like introversion or extroversion. The test is not a scientifically definitive measure of a city's personality but rather a tool to explore perceptions, strengths, weaknesses, and potential improvements. It is designed for both individual and group use, and it has been successfully applied in community planning and consultation. It provides a unique way to discuss cities, towns, or neighbourhoods by anthropomorphising them, allowing people to talk about these places as if they were individuals.

We asked our students to form groups and take the test for Tokyo. The results offer an engaging perspective on how young residents perceive the character of their city. From the viewpoint of our students, Tokyo might lack the excitement and innovation they crave. Some feel Tokyo is a bit too reserved and slow to make bold choices, leading to unresolved conflicts and a tendency to please everyone. Tokyo's strong sense of etiquette can make it seem a bit cold and unemotional, limiting the freedom for creative expression. The city's ambitions may come off as timid to young minds, and its focus on tradition could make it feel stuck in the past. Young people may perceive Tokyo as a place that does not dream big enough about the future and sometimes compromises values for practicality. While they acknowledge innovation, they see it often fading away without making a lasting impact. Overall, the city might benefit from a bit more excitement, a clearer vision for the future, and a willingness to embrace the bold ideas young minds have to offer.

Sustainability of Urban Design (Winter Term 2021–2022)

This class builds on the learning in "Psychology of Urban Design". Throughout the course of the term, students worked together to make sense of what exactly we need to do in order to sustain the built environment and make our cities liveable for future generations. Considering that by 2050, over two-thirds of the world population is projected to reside in cities (World Bank, 2023), the students were invited to inquire *who* will fall through the cracks, how a truly sustainable city can be planned, and how we should collectively deal with population growth, climate change, and mental health problems. Every week, students created new "working definitions" of sustainability, adapting them for the purpose of inquiry and discussion.

"Zoned Out"

We started the term with the notion of progressive individuals advocating for equity and inclusion in urban design in theory, but resisting concrete implementation of such policies "on the ground", especially in their own residential areas. Critics argue that some liberal-leaning residents, while supporting inclusive policies in principle, may oppose specific urban initiatives, like affordable housing or integrated neighbourhoods, when it directly affects their own communities (Harris, 2021). This perceived contradiction raises questions about the genuine commitment to equity and inclusion. We decided to explore this tension in class, emphasising the importance of aligning stated principles with tangible actions to achieve truly equitable urban development. We thus moved into the discussions of zoning and the specific case study of Los Angeles' Skid Row district.

Our objective was to cultivate a nuanced understanding among students regarding the intricate relationship between policy choices and social development. Initially, some students were unaware of the rationale and impact of specific policies (such as zoning) on urban design and the lived experience of specific communities. However, as we studied the Skid Row example and liberal hypocrisy (*The New York Times*, 2021) affecting American cities i.e., NIMBY ("not in my backyard"), the class began to recognise and identify the deliberate design choice of excluding humans from architectural decision-making, not just the invisible (homeless), but the hypervisible as well (teens).

Zoning in urban design refers to the practice of dividing land into designated areas for specific uses, such as residential, commercial, or industrial. Rigid zoning in many U.S. cities can lead to negative outcomes, fostering segregation, limiting housing affordability, and inhibiting mixed-use development (Gray, 2022; Talen, 2012; Hern, 2017). Zoning often results in spatial inefficiency, contributing to urban sprawl and excessive reliance on cars. Yet what we wanted to focus on was the inner city neighbourhoods negatively affected by years of neglect in terms of infrastructure planning and surrounded by extremely well-off and gentrifying areas that exacerbate the housing crisis as well as the mental health condition of the "undesirable population elements" (Harris, 2021). We wanted students to realise the extent to which policy affects specific urban areas, and to think of zoning policies that promote mixed-use spaces, enhance affordability, and foster vibrant, inclusive urban environments.

Although many communities across the political spectrum say they want to improve housing for all as a human right, some of those same advocates do not want this housing "in their backyard". In the case of California, where more people are moving in, but housing is unable to supplement the increased population, we asked students how they think this affects California's sustainability and in what terms. The class ended up bringing up single family zoning, regressive taxation and fragmentation of school districts as important factors that exacerbate the problem in Skid Row.

The activity revealed that affordable housing was the foremost issue being cited, and students made connections to neglected neighbourhoods in Japanese cities, too – for example, mentioning the *doya-gai* (slum-like districts and former hubs for day labourers) – in Japanese cities, such as Osaka's Nishinari, Yokohama's Kotobuki or Tokyo's San-ya. Students commented that the United States and Japan are, in their view, similar in terms of people often supporting the idea of affordable housing, yet being against the idea of providing affordable housing in the residential area where they live. There is no consensus on where exactly affordable housing should go. It is fair to say that communities that want to improve housing for all as a human right exhibit a change of heart when this housing becomes rolled out in their immediate neighbourhood, and the key voices being left out of the conversation are the very ones being affected the most – the unheard perspectives of the *invisibles*. The NIMBY crowd tends to be for affordable housing until these houses are in their area.

After watching the video prompt and carrying out a prolific discussion, students tackled the following question: "Can you think of other cities or neighbourhoods that have such buffer or 'containment' zones as Los Angeles? Please describe them and explain the connection(s)".

Some of the notable arguments included:

> In Osaka, there's a place that is similar to Skid Row called Nishinari and it's not what we imagine when we think about Osaka. The population of Nishinari consists of marginalised people such as Zainichi Koreans, Burakumin, homeless people, people with drug addiction and mental health issues. In the area, there's still many homeless who live in cardboard houses, people selling drugs in the street, and there is a red light district, and I think people who live in Nishinari also will be called "undesirable population elements" by the government, as well. Furthermore, Nishinari is getting gentrified because new buildings are getting built, causing the price of rent to increase and making it hard for people to afford living there.
>
> In Japan, we have places that are called ドヤ街「ドヤ」... which is a slang term in Japanese. It is made by reversing the word 「宿（やど）」 which stands for "housing". ドヤ in closest translation would mean "cheap housing". ドヤ街 is populated usually by men who are employed in manual labour. There is one famous place in Yokohama, called コトブキ, which is neglected by official authorities and may attract organised crime groups. In Kotobuki-cho, usually dock workers gather where there are many cheap lodging houses prepared for day labourers. While the area is known to house illegal gambling dens and numerous love hotels, it is clean and fairly hospitable compared to other places. Also, I think the name might be a little ironic, as 「寿」 means "long living".

Student connections between the American and Japanese contexts like these led to discussions on whether they would feel comfortable living in such neighbourhoods or their immediate surroundings. While our students come from privileged backgrounds, both absolute poverty levels and wealth inequality are lower in comparison to the United States. Despite lacking first-hand experience of life in neglected or under-privileged neighbourhoods, students nevertheless moved into discussions about the impact these might have on the sense of community and well-being for residents. They considered factors such as safety, access to resources, and social cohesion. Additionally, they reflected on the role of socioeconomic privilege in shaping perceptions and experiences of living in different neighbourhoods. The overall feedback on this activity was positive, and the students gradually moved from recognising to evaluating not just the socioeconomic, but also the psychological impacts of specific policies:

> I also enjoyed the assignment about Skid Row. I was able to identify the psychological border created in the district, which was caused by various policies . . .
> I learned about the use of street lighting and police presence and how these serve to divide and separate developed areas and underdeveloped areas.
> I also enjoyed the discussion about the usage of non-visual-borders in Skid Row. I understood that they can create an intentional, psychological, "us vs. them" ideology. If there was a solid, actual wall, people inside have a reason to fantasise what the outside would be like, and a reason to try and get over. However, when there isn't a wall, people are immediately hit by reality.

In our teaching sessions, we prioritise student interests, and during discussions on safety design techniques in urban environments, a topic often associated with hostile architecture, ethical considerations took centre stage. Students raised crucial questions about the ethical implications of shaping public spaces to influence user behaviour. As one student reflected, "urban design can be a form of oppression, such as hostile architecture", highlighting its negative impact on targeted groups, including the homeless and teenagers who face limitations on outdoor socialising ór engaging in activities like skateboarding or simply socialising in groups. This perspective underscores the varied ways in which hostile architecture affects different users, prompting critical examination of its consequences on urban inclusivity and well-being. The analogy between the invisible marginalised homeless populations and hypervisual teenagers, while not perfect, highlights design flaws in urban spaces. It draws attention to the limitations faced by teenagers due to hostile architecture and (potentially) zoning policies, emphasising the need for inclusive and equitable design practices. This comparison prompts a critical examination of urban design, urging considerations for the diverse needs of all community members.

"Out With the Old and In With the New!"

We continued the lessons with an unusual segue to a very different example of failed urban development projects. While Skid Row represented a historic and persisting issue of homelessness and poverty, our new topic – the relocation of capital cities – was about initiatives undertaken with intentions to spur development, which can lead to failed projects if not thoroughly planned and executed. In many ways, moving the capital did not solve existing issues in the cities that held that title before, but exacerbated (and "relocated") them. We believe that failed urbanism projects, whether in addressing housing issues or relocating capital cities, emphasise the importance of holistic, well-thought-out approaches to urban development.

We first introduced students to Indonesia's plans to move its capital from Jakarta to a new site in Kalimantan due to overpopulation and environmental issues. Our intention here was for students to explore the potential costs and benefits of the move, including social, political, economic, and environmental factors. They were then asked to research the following cities – Abuja (Nigeria), Naypyidaw (Myanmar), Brasilia (Brazil) and Nur-Sultan (Kazakhstan). It was a group assignment and each team had to list two reasons they thought were most important in the decision to move the capital and explain their choice, and evaluate the impact this decision had on the original/previous capital city.

Students reported back how Brazil moved its capital from Rio de Janeiro to Brasilia in 1960 to promote inland development. Kazakhstan shifted its capital from Almaty to Nur-Sultan in 1997, aiming to decentralise power and enhance economic growth. Nigeria proposed moving its capital from Lagos to Abuja in 1985 for strategic and administrative reasons, as well as due to overpopulation and climate change risks. Myanmar moved its capital from Yangon to Naypyidaw in 2005, citing a more central location and security concerns. Students noted how the impacts of capital relocation vary across geographical and cultural contexts. Potential benefits include regional development and administrative efficiency, while challenges include significant costs, social disruption, and potential geopolitical consequences. Evaluating the pros and cons involves weighing economic, social, and political factors, with the success of such endeavours contingent on effective planning and implementation.

Another observation – or rather, a question – from students was what happens to the old capitals? Students were genuinely interested in knowing whether or not the cause for the relocation would not only be addressed with the new capital, but the previous one that has already suffered under its burden. At this point, students understood that one indicator of good urban design is its sustainability, with a guiding thought to answer this question being, "Will this city meet the needs of my grandchildren?".

"The Shady Truth"

Another example of a case study that proved to be a transferable learning experience to the Tokyo metropolis context was the discussion of heatwaves and lack of urban greenery. Towards the end of the term, the class watched the video "How America's Hottest City is Trying to Cool Down" (Fong, 2021) about Phoenix, Arizona where trees are more than aesthetic elements. They function as essential infrastructure with services encompassing stormwater management, air filtration, carbon sequestration, and crucial cooling effects. Addressing the uneven distribution of trees across the city, Phoenix aims to achieve "tree equity" by 2030, acknowledging the role of trees in mitigating the rising frequency of extreme heat waves. American Forests (2022) defines tree equity as ensuring a sufficient tree presence in an area to enable everyone to access the health, climate, and economic advantages they offer. Trees play a crucial role in cooling and purifying the air, diminishing the likelihood of heat-related and respiratory ailments. Their contributions extend to filtering water pollution, mitigating flooding, lowering energy costs, sequestering carbon, and positively affecting mental well-being. Neighbourhoods with fewer trees typically expose residents to elevated levels of heat, pollution, and stress.

Phoenix's initiative responds to the disparities in greenery, particularly affecting lower-income neighbourhoods. The city is committed to using trees strategically to help reduce heat as part of its overall plan to combat high temperatures. Tokyo also struggles with heat and our students were eager to discuss solutions to improve the distribution of green spaces and ensure "tree equity" in the metropolis. They agreed that local assessments of green spaces, especially those aimed at identifying areas with inadequate tree coverage, would be the best starting point for mapping the problem and raising awareness of "tree equity". This would later involve collaborating with local authorities and communities to strategically plant trees and address environmental disparities within Tokyo.

Copenhagenize (Winter Term 2022–2023)

Why are the Danes consistently ranked as the happiest people in the world? One reason is because their cities combine sustainable action with people-centred urban design. Copenhagen, the capital of Denmark, is a pioneer when it comes to creating liveable, resilient, equitable, healthy, and sustainable spaces for all. In this class (third in our "series" of urban design courses), students took anthropology, sociology, urban planning, and common sense as the points of departure to explore what can make our living environment more eco-friendly as well as people-friendly. Over the course of the term, they learned about some common myths around cycling in the city, identifying best practices when it comes to "placemaking", public transport efficiency, and high standards of living. In a nutshell, they explored what it means (and what it takes) to "Copenhagenize" a city.

The idea for this particular class was inspired by the book *Copenhagenize: The Definitive Guide to Global Bicycle Urbanism* (2018) by Mikael Colville-Andersen, a seminal work that explores the bicycle's transformative role in urban design. The author is a prominent urban mobility expert who draws on Copenhagen's success as a bicycle-friendly city to advocate for a global shift towards prioritising bicycles as a sustainable and practical mode of urban transportation. Colville-Andersen delves into the historical, cultural, and economic factors that have made Copenhagen a model city for cycling. It provides insights and practical guidelines for cities worldwide to embrace bicycle urbanism and incorporate cycling into urban infrastructure.

Tokyo has been a city of cyclists for years, yet not because of the infrastructure or official urban policy, but rather because of its people. In Tokyo, the modal share of bike journeys, i.e. number of people using bicycles as their mode of transport within the overall transport usage, is about 15% (Copenhagenize Index, 2019). Although the share is over 60% in Copenhagen, Tokyo fares reasonably well in terms of absolute numbers of commuters because there are millions and millions of people on their bicycles, using their *mamachari* utility bikes to carry goods and children to the store, school, or train station.

The rationale to build an entire elective class on a single city was that Copenhagen truly promotes a vision of cities designed for people, where cycling becomes a central element in creating liveable, sustainable, and inclusive urban environments. Furthermore, the perceived greater transferability of Danish cycling infrastructure, notably in Copenhagen, compared to Dutch counterparts like Amsterdam, lies in its adaptability to diverse urban contexts and alignment with modern design principles. Denmark's innovative and safety-focused approach, coupled with a scalable and inclusive design, makes it appealing to cities seeking cutting-edge and universally applicable cycling solutions. While both countries offer valuable insights, the Danish model's adaptability and emphasis on safety contribute to its perceived broader applicability in various urban settings, including such densely populated and uniquely built cities as Tokyo or Osaka.

While the class initially provided case studies and examples of successful urban transformations, as the term progressed, students turned into designers themselves – applying their theoretical knowledge, personal experience, and creativity to make design proposals for safer street junctions, streets mixing commercial and residential purposes, and finally wider urban areas that encompass a wide range of design solutions.

Mapping Mobility: 25 Places in Your Life

The first assignment in class had to do with mapping our patterns of engagement with the urban environment, creating a list of no more than 25 places we most frequently visit. The activity was based on the study by Alessandretti et al. (2018), which explores human mobility patterns, revealing that despite virtually limitless choices, individuals tend to frequent the same 25 places on average. Analysing the trajectories of millions of mobile phone users, researchers find universalities in human behaviour across cultures, ages, and genders. This consistent limit in the number of frequented places suggests a deep-rooted aspect of human nature, challenging the notion of monotony in personal routines. The findings underscore how scientific advancements in studying human mobility contribute to understanding shared patterns within our diverse individual routines.

The follow-up to this activity was to critically examine why those 25 places are most important for us and what the design of their surrounding area is. The task revealed that the findings of Alessandretti et al. (2018) align and resonate with the GAPS philosophy, emphasising green, active, pro-social, and safe spaces in urban design. This understanding also links to concepts such as the 15-minute city, promoting proximity and accessibility to key locations. By critically examining the significance and design of these places, the assignment offers practical insights into creating inclusive, vibrant, and well-connected urban spaces, echoing considerations in sustainable urban development.

Street Designs

The next step was to delve into creative design ideas. Using the Kid-Sized Cities project, which serves as a platform to curate and share the design perspectives of children globally, we asked students to submit proposals for generic streetscapes using the web-based tools at https://cyklokoalicia. sk/kidsizedcities/scene/streetscape/. The Kid-Sized Cities initiative recognises children as insightful contributors to urban planning, leveraging their rationality and logic to envision improved streets, neighbourhoods, and cities. The goal of the project is to curate children's ideas globally, allowing them to express their thoughts through voices and visualisations, highlighting a shared understanding across diverse locations. The activity allowed us to witness the students' contagious enthusiasm and simple yet brilliant design thinking.

Before setting the assignment, we ventured outside the classroom once again to appraise the design of the street that our campus is on. Taking inspiration from Colville-Andersen once more, we asked students to consider if, based on their immediate surroundings where they spend most of their time in (during school hours), their city "fits" them. Colville-Andersen's daughter, Lulu-Sophia, questioned if her city fit her at the age of 4. Inspired by her profound observation, her father coined the term and philosophy. He pondered the notion of cities accommodating their residents and embarked on a mission to ensure everyone feels as if their city is designed for them. This concept evolved into a documentary series, emphasising the importance of cities being "life-sized" to meet the needs of every citizen. We decided to join the urban designer in his quest of refining ideas on creating urban environments that truly suit their inhabitants. The following images present the user interface and "menu" of design options that one can apply (Figure 7.11), as well as an example of a finished proposal (Figure 7.12).

Home   Play ▾

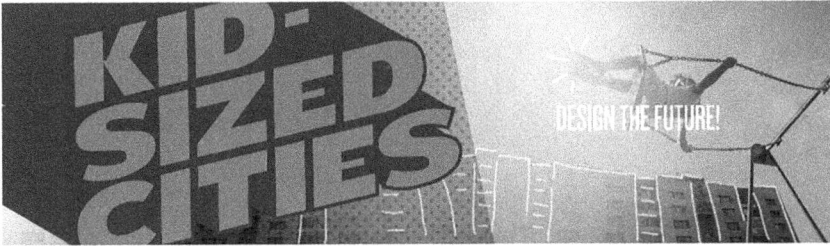

How would you make this city street into a "good street"? You can drag and drop elements into place. When you're finished, you can print it out and color it - and add stuff that we forgot.

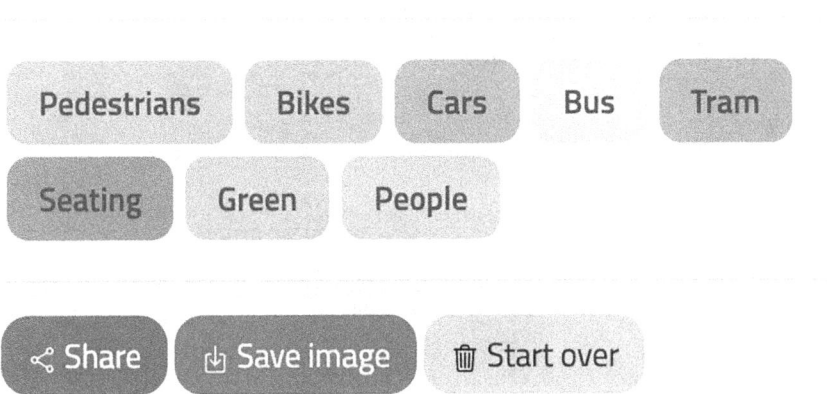

Pedestrians   Bikes   Cars   Bus   Tram

Seating   Green   People

≪ Share   ⬇ Save image   🗑 Start over

Figure 7.11 Kid-Sized Cities interface

Simple street designs were followed by assignments looking at safer intersections for different traffic users and, finally, the "arrogance of space" concept. Students recognised that although Tokyo is renowned for a robust and efficient public transit system and is also mindful of allocating space to cyclists and pedestrians, it still remains very car-centric. Elevated highways often negatively affect pedestrianised areas and cyclist spaces, creating physical and psychological barriers. These structures disrupt the open, human-scale atmosphere, diminishing the cohesiveness of pedestrian zones. Noise, air pollution, and visual intrusion contribute to a stressful environment, affecting residents' well-being. The disruption of natural urban flow and the imposing structures undermine community connection in pedestrianised areas, emphasising the need for thoughtful urban planning for the mental and physical health of city dwellers.

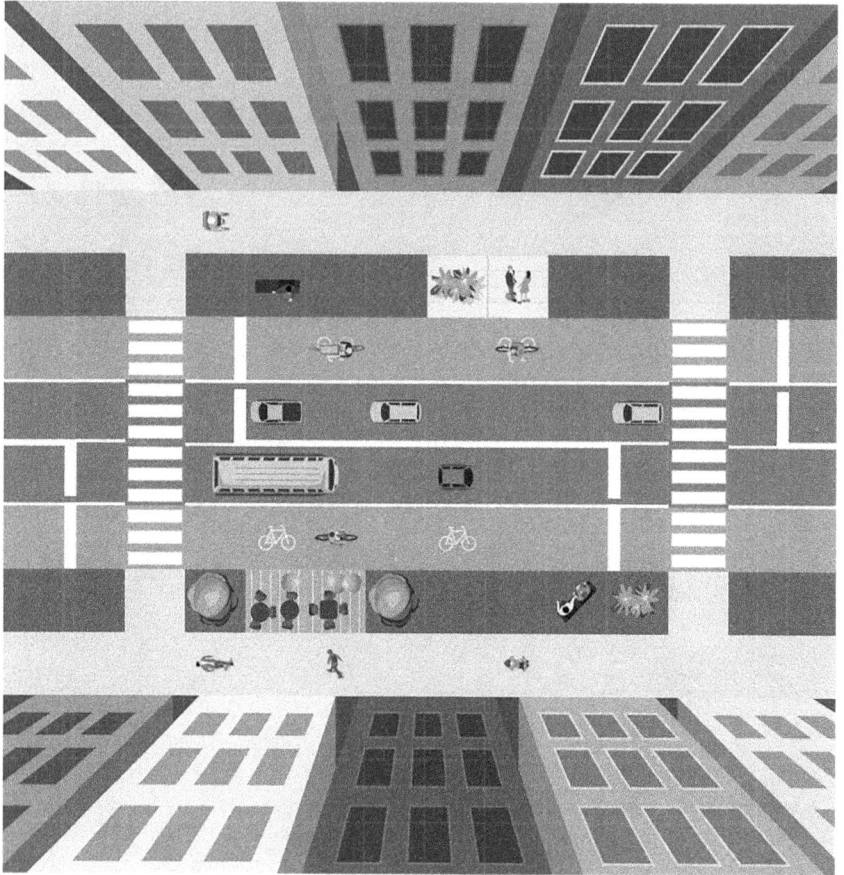

Figure 7.12 Kid-Sized Cities street design example

Arrogance of Space

It is here that the concept of the "arrogance of space" (Colville-Andersen, 2019) draws attention to the inefficiency and misuse of urban space for accommodating cars. The traditional allocation of city space for car-centric infrastructure, such as roads and parking, is not only unsustainable but also contributes to the degradation of urban quality of life. Cities around the world, from Bogota to Paris to Melbourne, are rethinking their urban planning and design, advocating for reallocating space to prioritise pedestrians, cyclists, and public transport. Colville-Andersen emphasises the arrogance embedded in dedicating vast areas to cars, which often leads to congestion, pollution, and a lack of social interaction. Young people around the world can be the agents of change and encourage the paradigm shift towards more people-centric urban design, whereby streets and public spaces are reclaimed for the well-being of the community rather than prioritising the convenience of automobiles. This is reflected in the following students' views:

> In my opinion, urban design should focus on allocating elements equally. Currently, a lot of cities are prioritising certain things and underestimating others, which is leading to an unequal allocation. For example, cars are often the most prioritised element in cities and this is affecting cities to allocate spaces for cars more than enough. On the other hand, green spaces are underestimated and there are very few green areas in cities.
>
> In my opinion, the next urban design class should include a thorough traffic flow analysis, as well as an examination and comparison of the cost-effectiveness and practicality of potential reconstruction projects.

For the main piece of assessment in class, students were given a choice of four cities with specific areas to create a new design for, all meant to re-allocate the space more evenly for people – whether they walk, cycle, drive, or take public transport. For areas full of concrete, they were encouraged to consider adding green spaces. The specific location in each city is a busy urban area – each one has been subject to either traffic redesign or full revamp proposals for a long time now. Students were given the freedom to go for drastic changes or very minimal ones – if they ended up not changing much, they were asked to be prepared to explain why the current arrangement works. The four "arrogant" spaces to "fix" were Constitution Square in Warsaw, the Elephant & Castle roundabout in London, Place de la Concorde in Paris and the Tverskaya Zastava square in Moscow. Students used Google Maps to explore their chosen areas – zooming in, rotating the screen, switching between 2D and 3D views, etc. They could also view the specific location on Google Earth. The next step entailed uploading screenshots or photographs of the urban areas under consideration to the Arrogance of Space mapping tool online at https://cyklokoalicia.sk/arrogance/. This allowed students to divide their image(s) into squares and label them to calculate how much space is used by cars, pedestrians, etc. The final step was to sketch ideas for redesigning the area and present to the rest of the class using the "marketplace" format.

Students' redesign proposals showcased a sound understanding of the curricular criteria (i.e., "I can assess the impact of developments in transport infrastructure in a selected area and can contribute to a discussion on the development of sustainable systems"), as well as various innovative features to enhance urban spaces. Some examples include:

- Pedestrianised zones: Many designs incorporated expanded sidewalks, pedestrian-only areas, and enhanced crosswalks to prioritise walking.
- Cycling infrastructure: Students introduced dedicated bike lanes, often uni-directional (Copenhagen-style); bike-sharing stations; and bicycle parking to promote cycling as a sustainable mode of transport.

- Green spaces: Student proposals focused on integrating more greenery, such as parks, trees, and even urban meadows, to improve air quality and create aesthetically pleasing environments.
- Public transit hubs: Designs often featured improved public transport facilities, including efficient bus stops, tram stations, and subway entrances, encouraging the use of mass transit.
- Traffic measures: To enhance safety, some students suggested traffic-calming measures like speed bumps, pedestrian islands, and roundabouts.
- Smart urban furniture: Incorporation of smart benches, interactive displays, and modern street furniture aimed at providing utility and enhancing the overall experience.

These examples illustrate the students' thoughtful consideration of diverse elements, reflecting their commitment to creating inclusive, sustainable, and vibrant urban spaces for all. The class ended just before Spring Break and we could not think of a better way of saying goodbye than going for a walk with a new set of lenses to use in appraising our urban environment. We went to Ike Sun Park in Ikebukuro to study it in detail. The purpose was to explore the area and assess whether Ike Sun Park "worked" – in terms of the traffic arrangements in its immediate surroundings, in terms of its aesthetic beauty as a green space, and in terms of the purpose it serves in urban design. As we walked there, students reflected about what they learned about "Copenhagenizing" a space. They looked around carefully to determine whether this part of Tokyo had an easy-to-follow, consistent bicycle infrastructure; whether the place was "life-sized"; and whether there were any spaces that could be considered "arrogant". The field trip revealed what some of the next steps in student learning could be, with many voices calling for a deep dive into the design of multi-purpose, disaster-ready spaces in Tokyo.

Middle School Social Studies Curriculum

At the start of the academic year 2023–2024, we modified certain elements of our previous high school courses to present intricate urban design concepts in an accessible and engaging manner for a younger audience. The lessons and activities discussed in this section were designed and delivered to "Upper School" students, i.e. the junior high phase equivalent with students aged 12–15. Our school adopts a multi-age philosophy, however, meaning that grades 7–9 are grouped together. Recognising that this class was mandatory and drawing insights from past experiences where certain concepts were culturally and linguistically "alien" or distant to students, we aimed to familiarise them with the crucial link between urban environments and personal well-being. Tailored for junior high school students, the initial term focused on unravelling the symbiotic connection between urban design and mental health.

The overarching learning intention for the year was to embark on a journey to forge stronger connections with the buildings and spaces that shape our surroundings. Our collective mission was to unravel the key actions required to uphold the integrity of the built environment, ensuring that our cities remain "liveable" for generations to come. We delved into history, geography, urban planning, and common sense, using them as our starting points to investigate ways to transform our living spaces into environments that are both eco-friendly and people-friendly. The 100-year anniversary of the Great Kanto Earthquake was our first topic. It focused on safety standards and design principles in Japan. While this devastating seismic event provides crucial lessons for contemporary urban design education (by highlighting the imperative of integrating earthquake-resistant principles into architectural planning), junior high students have displaced reluctance due to a perceived disconnect from their immediate experiences and the historical context. Engaging them requires framing the lessons in a relatable manner, emphasising the enduring relevance of safety standards, and fostering an understanding of how past tragedies inform present-day strategies for creating resilient, secure urban environments. We decided to come back to this later in the year, opting to instead focus on urban design and emotional well-being.

Classes 1 and 2: Discovering Urban Design and Mental Health

Our introductory session aimed to familiarise junior high school students with the fundamental concepts of urban design and its potential impact on mental health. We had to spend a few more lessons than planned on clarifying to students the difference between urban design and urban planning, and that a singular building is not an example of urban design. Grounded in the experiences of their local community, students engaged in individual and group activities, sharing thoughts on what makes spaces special to them.

Building on the foundation laid in the first few weeks of class, we eased them in with a TED talk (Hood, 2018). We understand the irony of using a TED talk to ease students into a lesson and an emotion mapping activity. By using relatable examples and simplified discussions, students were encouraged to express their thoughts on how urban spaces influence emotional and mental well-being. It was at this point when we also had to introduce new vocabulary and ideas to students. When asked to emotional map, they had difficulty distinguishing what constitutes an emotion and a situation. An example of this is when shown a picture of a heavily (foot) trafficked area and asked how they felt, they would often respond with adjectives such as "busy" when in fact they meant overwhelmed. Despite the initial challenges in distinguishing between emotions and situations, the students gradually developed a more nuanced understanding through continued exploration of urban design concepts and vocabulary.

Classes 3 and 4: Infusing Art and Culture

In these classes, students were prompted to reinfuse life into seemingly mundane spaces, preparing the groundwork for understanding how art and cultural elements can contribute to urban design. The free and intuitive 3D modelling tools available as part of the "SketchUp for Schools" project provided a platform for students to envision and articulate their ideal urban spaces with an emphasis on artistic and cultural aspects. Again, it was evident that some students were still having difficulty understanding that a singular building does not constitute an urban area. One way we adapted to this challenge, and with the hopes of encouraging rather than deterring, we expanded the project to allow students to continue adding spaces based on new concepts and lessons that have been and will be learned in the future. We circled back to the "life between buildings" idea (Gehl, 2010) – that is, the emphasis on the significance of public spaces in connecting people and enhancing the overall quality of urban life. We showed students examples of urban rejuvenation through art, like the decorated courtyards in Dresden's Kunsthofpassage or Poznań's "Green Symphony" mural, which showcases art's transformative role in urban spaces. These installations not only enhance aesthetics but also foster community and repurpose neglected areas into vibrant hubs, aligning with Gehl's vision for dynamic, people-centric cityscapes that prioritise well-being.

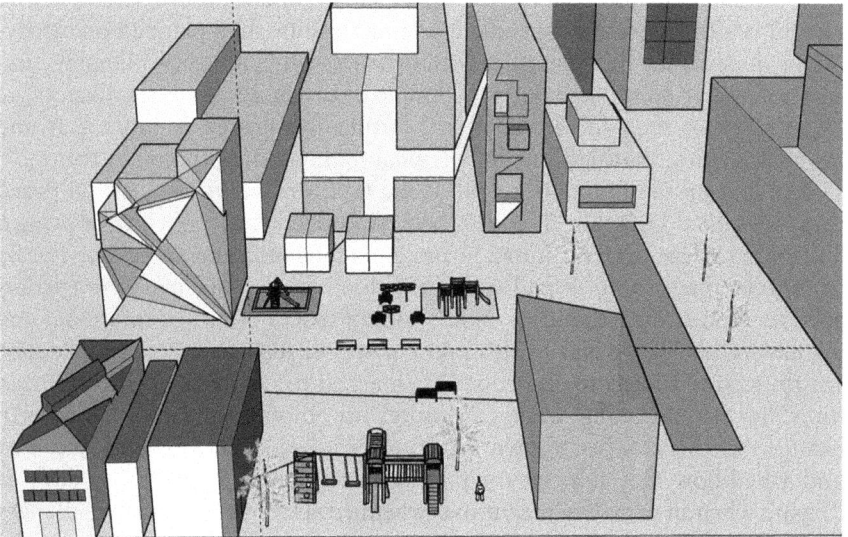

Figure 7.13 SketchUp example 1. This particular design focused on creating art and cultural spaces for all games

Figure 7.14 SketchUp example 2. This one focused on walkability

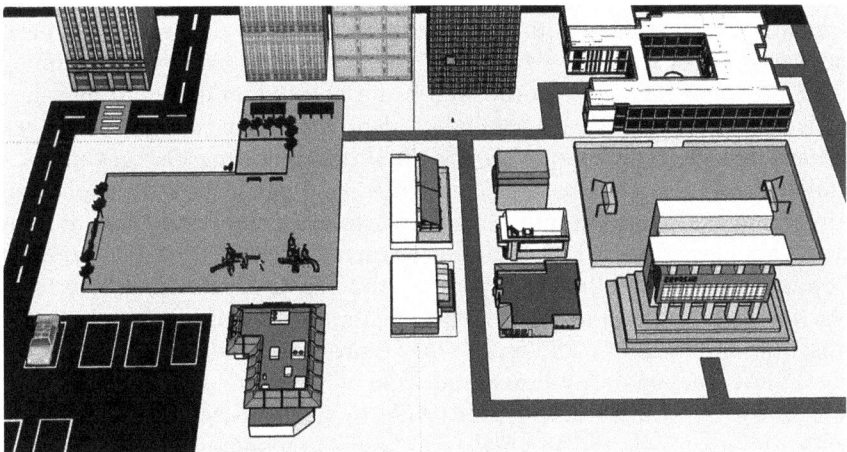

Figure 7.15 SketchUp example 3

Class 5: Tuning Into Soundscapes

Acknowledging the unique perspectives of junior high school students, the fifth class focused on the auditory component of urban design. Through a tailored sensory mapping activity for soundscapes, students gained insights into the impact of different auditory environments on the body and mental well-being. We tasked the students with taking five one-minute samples of different soundscapes in their environment; the students were to bring those samples back and analyse the recordings together as a class to evaluate how sounds shape the whole picture of their urban area.

This activity links with Gordon Hempton's concept of "One Square Inch of Silence" (2009) and his work on preserving natural soundscapes, which have relevance to urban areas by encouraging awareness of the acoustic environment within cities. While urban spaces are inherently noisy due to human activities and infrastructure, analysing soundscapes and noise levels prompts reflection on the importance of incorporating elements of natural sound into urban planning. It advocates for creating pockets of relative quiet or incorporating green spaces that allow residents to experience a more balanced and natural acoustic environment amid the urban hustle. Hempton's ideas encourage a holistic approach to urban design, acknowledging the significance of soundscapes in enhancing overall well-being and creating more harmonious urban environments. In our exploration of the impact of urban unsustainability and soundscapes, we explored through three key classroom activities that shed light on the historical, evolving, and contemporary dimensions of the urban sound environments.

The first activity, *Researching Historical Sounds*, delved into the transformation of sounds over time. Students looked into archives of the past, uncovering what sounds were prevalent in different eras and examining their potential influence on the well-being of communities. Through historical lens, it became (hopefully) evident that soundscapes in a given urban area have undergone significant shifts, reflecting changes in societal structures, norms, economic activities, and particularly technological advancements. This naturally lended itself to a seamless transition into our next talking point on how soundscapes have evolved. The *Evolution of Soundscapes* activity further expanded students' understanding by tracing the historical changes in soundscapes. Examining factors such as industrialisation, urbanisation, and technological progress, students explored how these shifts shaped the auditory landscape of urban environments. The rising hum of machinery during industrialisation, the increasing traffic noise with urban sprawl, and the pervasive hum of technology all contributed to an intricate tapestry of urban soundscapes and the hand they play in human behaviour, like aversion and sleep cycles. This exploration highlighted the intricate dance between societal developments and the evolving soundscape, raising questions about the potential repercussions on mental and physical health.

As we transitioned to the contemporary realm, the *Quiet Spaces in School* activity provided a reflective perspective on the implications of an environment's soundscape for mental health and behaviour in an educational setting. The class engaged in surveys and discussions investigating how the availability of quiet spaces could influence students' stress levels. The findings demonstrated the importance of tranquil environments in mitigating stress, enhancing concentration, and promoting overall well-being. This microcosmic exploration within the school environment underscored the broader significance of creating peaceful urban areas that cater to the mental health needs of their inhabitants. It is important to note here that the school, as most schools in Tokyo, does not have a campus of its own; instead, we are housed in buildings in a small commercial area where the city's soundscape can often overtake a classroom.

Our classroom activities underscored the intricate relationship between urban unsustainability and the creation of new soundscapes with adverse effects on both the body and mental health. From historical sound influences to the evolving dynamics of urbanisation, and the microcosmic impact within school spaces, the class explored varying vistas that collectively emphasise the sustainability of urban planning that prioritises the auditory well-being of its residents.

Nurturing Understanding in Urban Design and Mental Health

The second part of the lesson plan aimed to deepen students' comprehension of specific urban design elements and their potential effects on mental health. Utilising mini-lectures, group research projects, and presentations, the curriculum adapted complex concepts to suit the cognitive level of junior high school students, exploring key aspects such as green spaces, walkability, safety, and access to amenities – activities we have discussed prior with the previous classes.

Tailored for the unique learning needs of the 12–15 age group, the lesson plan serves as an educational bridge, nurturing an early appreciation for the interconnectedness of urban design and individual well-being. In one student's reflection on their learning, they stated:

> Before learning about urban design, I only knew about aspects like road planning, which is what I see in Tokyo every day and what I do in city planning games. Now, after learning about urban design, I discovered many new concepts like walkability, mental effects, sounds, that completely change my conception regarding city planning.

Or as other students wrote (translated from Japanese):

> Before I knew about urban design, I only saw the city of Tokyo when I looked at the city, and I could only see one side of it. For example,

when I went to a city park before knowing about it, I only thought simple things like "This is a nice park". However, now that I know about urban design through school, when I go to a park, I can think about what impact this park has on people and why the park was built. For example, if there are trees and plants in this park, it will likely reduce stress and improve the quality of life for people. So, this park was built for that reason . . .

Before this semester, I didn't know anything about urban planning. However, after learning about it this semester in social studies, I became interested in urban planning and design. Also, by knowing about urban planning, I felt that it takes time to plan the city and also think about the design that can calm people down and relax them. The most interesting learning activity this semester was designing a city with SketchUp for Schools. Also, the activity of making an uninteresting design interesting was fun. I think I actively worked on my homework this semester. I want to apply what I learned this semester to observe the design of the city when travelling to various places, and I want to think about how to improve uninteresting designs if I find any.

Before this semester, I didn't think deeply about how factors like building height and noise levels could significantly affect people's minds. However, now I can think about opportunities related to sunlight and nature.

These quotes share a common theme of transformation in perspective and awareness through the study of urban design and planning. The students express how their understanding has evolved from almost none or superficial observations to a nuanced appreciation of how urban elements affect people's lives. The newfound ability to analyse aspects like park design, city planning, soundscapes, and environmental factors reflects a deeper engagement with the built environment.

Next Steps

Based on student feedback and interests, as well as unused ideas we have already prepared, we have a range of potential topics and new directions for study and discussion in future classes, such as the intersection of technology and urban development and the integration of those technologies into city planning; technologies that can be used to enhance the quality of life for who resides in the area. Another array of lessons that can be built around the "15-minute cities" and "smart cities" ideas, the use of data analytics, the implementation of smart infrastructure, and the implications for governance, privacy, and social equity (Montgomery, 2015). We are also interested in delving deeper into the move of capitals and shifting coastlines which lends itself to looking into flood or sea level management, alert systems, or more specifically – the "sponge city" concept (Shi et al., 2023). We can even revisit previous lessons with a new cohort of students about planned but failed utopian cities, and why they were so radical at the time, like *Metropolis* – city of tessellating hexagons on Niagara Falls where money would "pass into the oblivion of an ignorant age" (Heal, 2019).

We concur with Landry and Murray (2017) that embracing psychology in urban discourse will be crucial for addressing the fragility of cities, global challenges, and behaviour change in an ever-changing world. We hope that this chapter will serve as a source of questions to explore with young learners and stimulate discussion, debate, and collaboration – within schools and beyond – in order to address the challenges of rapidly changing urban landscapes and climate patterns. In the context of an international school in Tokyo, Japan, our exploration of urban design will continue to encompass key themes such as sustainable city planning, youth engagement in urban development, and the critical role of green spaces. Addressing climate change and fostering sustainability align with global educational priorities, making it essential for students to comprehend the broader implications of urban growth. As mentioned before, these concerns led to student projects that centres around the sustainability of infrastructure as a proactive approach. The engagement of young minds in urban design, as emphasised by initiatives like Kid-Sized Cities, brings valuable perspectives to the conversation, echoing the importance of inclusive planning, especially when some have already experienced first-hand how unsustainable urban planning and design can affect a nation when having to move capitals. Tokyo's urban challenges, like congestion and limited green spaces, necessitate a strategic approach, aligning with concepts such as the 15-minute city and Tokyo's impending "Day Zero" (Schapitl, 2018). Integrating these urban design principles into the school curriculum and collaborating with local communities can empower students to contribute actively to Tokyo's sustainable urban future, fostering a sense of responsibility, urban empathy, and global citizenship.

References

Alberti, M. (1999). Urbanization and the ecology of wildlife. In J. McDonnell & S. Pickett (Eds.), *The Ecology of Urban Environments*. Springer, pp. 183–207.
Alda, A. (Host). (2023, May 23). Daniel Libeskind: His buildings speak for themselves [Audio podcast episode]. *Clear + Vivid*. https://clear-vivid-with-alan-alda.simplecast.com/episodes/daniel-libeskind-his-buildings-speak-for-themselves
Alessandretti, L., Sapiezynski, P., & Sekara, V. et al. (2018). Evidence for a conserved quantity in human mobility. *Nature Human Behaviour, 2*, 485–491. https://doi.org/10.1038/s41562-018-0364-x
American Forests. (2022). *Tree Equity*. https://www.americanforests.org/our-programs/tree-equity/
Beatley, T. (2011). *Green Cities of Europe*. Washington, DC: Island Press.
Berman, M. G., Jonides, J., & Kaplan, S. (2008). The cognitive benefits of interacting with nature. *Psychological Science, 19*(12), 1207–1212.

Burchell, R. W., Shad, J., Gualtieri, G., & Downs, A. (2002). *The Costs of Sprawl – Revisited. Transit Cooperative Research Program Report 74*. Washington, DC: Transportation Research Board.

Coley, R. L., Kuo, F. E., & Sullivan, W. C. (1997). Where does community grow? The social context created by nature in urban public housing. *Environment and Behavior, 29*(4), 468–494.

Colville-Andersen, M. (2018). *Copenhagenize. The Definitive Guide to Global Bicycle Urbanism*. Washington, DC: Island Press.

Colville-Andersen, M. (2019). *The Arrogance of Space*. https://colvilleandersen. medium.com/the-arrogance-of-space-93a7419b0278

Copenhagenize Index. (2019). *The 2019 Copenhagenize Index of Bicycle-Friendly Cities*. https://copenhagenizeindex.eu/the-index

Engeström, Y. (1987). *Learning by Expanding: An Activity-theoretical Approach to Developmental Research*. Helsinki: Orienta-Konsultit.

Fong, J. (2021). *How America's Hottest City Is Trying to Cool Down*. https://www. vox.com/videos/2021/9/20/22683888/sonoran-desert-phoenix-tree-equity

Gehl, J. (2010). *Cities for People*. Washington, DC: Island Press.

Grant, M., & Perchoux, C. (2019). Urban sprawl. In M. J. Green (Ed.), *The SAGE Encyclopaedia of Landscapes of the Americas*. Sage, Vol. 4, pp. 1463–1467.

Gray, M. N. (2022). *Arbitrary Lines: How Zoning Broke the American City and How to Fix It*. Washington, DC: Island Press.

Harris, J. (2021). *Skid Row, Explained*. https://www.youtube.com/watch?v=rKo8 Sv99MkM

Heal, A. (2019). *How Gillette's Founder Dreamed of a Car-free, Moneyless Metropolis*. https://www.theguardian.com/cities/2019/jan/24/how-gillettes-founder-dreamed-of-a-car-free-moneyless-metropolis

Hempton, G. (2009). *One Square Inch of Silence*. New York: Free Press.

Hern, M. (2017). *What the City Is for. Remaking the Politics of Displacement*. Cambridge, MA: The MIT Press.

Hood, W. (2018). *How Urban Spaces Can Preserve History and Build Community*. https://www.youtube.com/watch?v=762c6pFpoqg

Kaplan, S. (1995). The restorative benefits of nature: Toward an integrative framework. *Journal of Environmental Psychology, 15*(3), 169–182.

Kennedy, C., et al. (2016). Compact cities with well-connected green spaces can reduce greenhouse gas emissions. *Nature Climate Change, 7*(5), 423–427. https:// doi.org/10.1038/nclimate3243

Landry, C., & Murray, C. (2017). *Psychology and the City: The Hidden Dimension*. Stroud, Gloucestershire: Comedia.

Lees, L. (2012). Gentrification and social mixing: Towards an inclusive urban renaissance? *Urban Studies, 49*(10), 2071–2087.

London First. (2017). *Not Just a Pretty Face: A New Agenda for Shaping London*. https://www.businessldn.co.uk/sites/default/files/documents/2018-05/Not-just-a-pretty-place.pdf

McGranahan, G., Marcotullio, P., Bai, X., Balk, D., Braga, T., Douglas, I., Elqvist, T., Rees, W., Satterthwaite, D., Songsore, J., & Zlotnik, H. (2005). *Chapter 27 Urban Systems*. https://www.iied.org/g00505

Montgomery, C. (2015). *Happy City. Transforming Our Lives Through Urban Design*. London: Penguin – Random House.

The New York Times. (2021). Liberal hypocrisy is fueling American inequality. Here's How. | NYT opinion. https://www.youtube.com/watch?v=hNDgcjVGHIw

O'Sullivan, F., & Bliss, L. (2020). The 15 minute city – no cars required – is urban planning's new Utopia. *Bloomberg Businessweek*, 12 November. https:// www.bloomberg.com/news/features/2020-11-12/paris-s-15-minute-city-could-be-coming-to-an-urban-area-near-you

Pickett, S. T., Cadenasso, M. L., Grove, J. M., Nilon, C. H., Pouyat, R. V., Zipperer, W. C., & Costanza, R. (2001). Urban ecological systems: Linking terrestrial ecological, physical, and socioeconomic components of metropolitan areas. *Annual Review of Ecology and Systematics*, *32*, 127–157.

Romero-Lankao, P., McPhearson, T., Davidson, D. J., & Bettez, N. (2012). Governance of the coupled human–environmental system: The case of climate change adaptation and cities. *Current Opinion in Environmental Sustainability*, *4*(4), 465–472.

Sallis, J. F., Floyd, M. F., Rodríguez, D. A., & Saelens, B. E. (2012). Role of built environments in physical activity, obesity, and cardiovascular disease. *Circulation*, *125*(5), 729–737.

Schapitl, L. (2018). By 2040, most of the world won't have enough water to meet demand year-round. *Vox.* 12 September. https://www.vox.com/2018/9/12/17842888/world-water-crisis-day-zero-explained-netflix

Shi, C., Miao, X., Xu, T., Gao, W., Liu, G., Li, S., Lin, Y., Wei, X., & Liu, H. (2023). Promoting sponge city construction through rainwater trading: An evolutionary game theory-based analysis. *Water*, *15*(4), 771. https://doi.org/10.3390/w15040771

Talen, E. (2012). *City Rules: How Regulations Affect Urban Form*. Washington, DC: Island Press.

Tokyo Metropolitan Government. (2023). *Population of Tokyo*. https://www.metro.tokyo.lg.jp/english/about/history/history03.html

Urban Psyche. (2020). *City Personality Test*. https://test.urbanpsyche.org/

Useem, J., Useem, R. H., & Donoghue, J. (1963). Men in the middle of the third culture: The roles of American and non-Western people in cross-cultural administration. *Human Organization*, *22*, 169–179.

Vygotsky, L. (1965). *Thought and Language*. Cambridge: MIT Press.

Ward Thompson, C., Roe, J., Aspinall, P., Mitchell, R., Clow, A., & Miller, D. (2012). More green space is linked to less stress in deprived communities: Evidence from salivary cortisol patterns. *Landscape and Urban Planning*, *105*(3), 221–229.

Wolch, J. R., Byrne, J., & Newell, J. P. (2014). Urban green space, public health, and environmental justice: The challenge of making cities "just green enough". *Landscape and Urban Planning*, *125*, 234–244.

World Bank. (2023). *Urban Development: Overview*. https://www.worldbank.org/en/topic/urbandevelopment/overview

8 Instilling a Sense of Wonder in Elementary and High School Science Students

Wonder-Infused Pedagogy Makes a Difference

Betty Trummel

Elementary school teachers often experience a disconnect with science, feeling unprepared to dive deep and engage students in essential topics and content. Little university study on methods of teaching elementary science often leads to very routine instruction, lacking deeper understanding and inquiry, and proving to be rather unexceptional. Even a well-trained high school educator, who delivers plenty of content and methods, can in fact be less than enthusiastic or creative at times.

But, what about students who experience a committed and exceptional elementary or high school science educator? The preparations and experiences of those educators who are both skilled and enthusiastic about science education can make all the difference. Their excitement can be contagious and motivating. The sense of wonder they instill in their students can aid learners in becoming critical thinkers, help them find joy in learning, and keep them interested in science for the remaining years of their education and as a possible career.

For many students, having a lackluster science teacher in the early years can terminate their interest and excitement for this subject. Betty's classroom (and life) was forever changed after her first science research experience at McMurdo Station in Antarctica in 1998. Her 9- and 10-year-old students were in a classroom infused with a depth and breadth of knowledge and activities they would not have had without her first-hand experiences. Two additional research projects in Antarctica and numerous trips to work with teachers in the Arctic helped give her the opportunity to provide a unique learning environment for her students.

This chapter will provide evidence of teaching practices and two classroom teachers who have made a difference. Examples of how these two educators used polar education and sustainability as key topics, particularly based on their own experiences in the polar regions, will be the focus of the case studies.

DOI: 10.4324/9781003486961-8

Case Study 1: An Elementary Classroom Filled With a Sense of Wonder

A sense of wonder in life and learning – Betty didn't give that much thought in her early years. She grew up on a small "hobby farm" in New Jersey, with a deep connection and affection for nature and being outdoors. As many teens or young adults do, Betty became disconnected temporarily from nature, and although she still loved the outdoors, she seldom explored and had adventures in learning like she had in her youth. At no point in Betty's early years and high school education did she cross paths with an inspiring science educator. She left grade 12 with little to no enthusiasm for science. Betty had always wanted to teach elementary school, but teaching science was not a big part of the plan. This uninspiring lack of connection with science persisted throughout her university training, as well.

A Transformational and Intergenerational Outdoor Education Success Story

Fast forward to Betty's early years of being an elementary educator. She felt like her college courses did not prepare her to teach science as part of a self-contained classroom. Yearning for some professional development in this subject area, she found a summer program called "Conservation Summits" offered by the National Wildlife Federation (NWF). Back in 1983, Betty embarked on what would be one of the seminal experiences that would influence her pedagogy, teaching career, students, and ultimately her family, and give her life in education major purpose.

Involving families and individuals from infants to octogenarians, the Conservation Summits program (started in 1970) would bring large groups of like-minded people together in a spectacular outdoor setting for a week of outdoor learning experiences. At that time, each summer the NWF would host Conservation Summits in several locations around the United States and, on a few occasions, in Canada. Great care was given to site selection, always finding places that would capture the wild spirit of nature, offer opportunities for learning about the flora and fauna, and engage attendees in the natural and local history of the location.

Betty's first Summit, in 1983, was held at the YMCA of the Rockies just outside of breathtaking Rocky Mountain National Park in Estes Park, Colorado. This geographical location became a pivotal location in her future teaching endeavors, and also for her family.

Engaging in classes on such topics as butterflies, alpine flowers, wildlife ecology, wilderness medicine, backpacking, and being a nature "creep" started to bring back the sense of wonder Betty had felt as a child. By becoming a nature creep, participants slowed down their pace and moved at ground level to find the intricate details of nature right at their feet. Summiteers felt energized and inspired, and Betty was ready to start a new school year and transfer this knowledge and enthusiasm to her students and friends, as well. Betty's classroom was transformed into a place of wonder, and her teaching methods had been transformed. Enthusiasm for teaching science had grown exponentially, as did the excitement of the fourth grade students Betty taught. The learning environment was vastly different from what it had been the prior year.

What Betty had not counted on was her own enthusiasm for getting involved in teaching through this program. Since 1983, she has taught at well over 45 Summits. For the first 20 years, she directed the Big Backyard Pre-school Program and taught children ages 3–4. The past 18 years took Betty in another direction. She transitioned to being a naturalist, leading "rambles," which are hikes that really hone in on the natural features, plants, and animals that can be found along trails in a multitude of ecosystems. Much like the original nature creep, participants taking part in nature rambles are stopping to creep along or take time in nature, similar to what Betty was taught almost 40 years ago. This is the hallmark of a natural ramble.

The extremely talented educators and science experts Betty met became the impetus for a myriad of projects and the force that drove her teaching career and desire to complete a Master's Degree in outdoor education in 1991. Many of those Summit teaching colleagues, regardless of where they lived, ended up being guest speakers in her fourth grade classroom or for whole school assemblies. Biologists, geologists, wilderness photographers, authors, experts in nature journaling, and many others became part of Betty's regular classroom routine. What an impact this had on hundreds – if not thousands – of students in the 30 years of teaching that followed Betty's first Summit experience (Family Nature Summits, 2023).

Summits were a catalyst for Betty as a learner and educator. New units of study, engaging and in-depth projects, and connecting students to the natural world became the norm. A classroom library full of both non-fiction and fictional science-themed books was freely available to all. Teaching about energy, recycling, wildlife, national parks, wildflowers, explorers, naturalists, and unique environments became a part of every single day in "Mrs. Trummel's fourth grade classroom." Her classroom was full of artifacts from the natural world and was a mini-museum of sorts. Betty was the whacky science lady, whose students built a kelp forest hanging from the ceiling, pushed back their desks and brought in sleeping bags to "camp" in a national park, hosted fabulous science events and shared them with peers and parents, and whose students still remember their fourth grade year because the learning was impactful and *fun*!

As a result of this infusion of wonder into Betty's life, and subsequently her classroom, she could see her students opening their eyes to learning in ways she had not experienced as a younger person. Science, reading, and other subject areas were suddenly connected in her classroom, not seen as separate areas of instruction, but more a blend throughout each day. A passion for science, technology, and literacy grew with each passing year as Betty was developing her pedagogies and finding her way in education.

Betty's application for the Presidential Science Awards in 1996, as well as other projects she was engaged with at this time, meant she was continually examining and reflecting on pedagogies and practices while striving to improve her own content knowledge and that of her students. Betty was on a mission to find what would inspire her to keep learning and to continue to create that sense of wonder for students. Perceptions and attitudes toward science were interesting to her, and as a result, Betty started off the science curriculum each year by asking students to "draw a picture of a scientist at work." That was the only instruction, and with more than 20 years of pictures, the results have told the story.

Figure 8.1 Beginning of a fourth grade year

Figure 8.2 Middle of the school year

Figure 8.3 and *Figure 8.4* Depicting placement of flags on a sediment core in Antarctica

Figure 8.3 and *Figure 8.4* Continued

Figure 8.5 Diving for research in a kelp forest

Most drawings depicted men with crazy hair working in laboratories, mixing chemicals, and wearing lab coats, glasses, and pocket protectors. Betty was out to prove that having an inspiring and motivational educator could make a difference in that perception. With each passing unit of instruction and unique project, inquiry-based lesson, or special scientist connection, students slowly changed their ideas about what science encompasses and who can be a scientist. When it came time to draw a picture of a scientist at work at the end of the school year, the pictures were tremendously different from when students had entered fourth grade. There were women (and men) working indoors and outdoors, and in all areas of science and discovery.

It was very obvious that is was not just a matter of curriculum or of a school district or state's instructional guidelines, but how as a teacher Betty was able to improve and refine her skills to have a lasting impact on her students. Actions transferred directly to students – it mattered! This demonstrated that what and how educators teach and their own personal pedagogy is critical (Kelly, 2018).

At a recent event in Crystal Lake, Illinois, the town where Betty taught for 27 years, she had the opportunity to reconnect with six former students. Every single one of them talked about how impactful their fourth grade year with Mrs. Trummel had been. All of them talked about how exciting, interesting, and interactive that classroom had been and how much they loved fourth grade. They gave specific examples of what projects and activities they both remembered and loved. They were shouting out "the classroom camp-in," "building the kelp forest hanging from our classroom," "the connection to Antarctica and the science taking place there," "the energy fair," "following the Iditarod sled dog race in Alaska," "orienteering field trips," "hiking through Volo Bog State Natural Area and all of the other outdoor science field trips," and "all of the guest speakers" they were introduced to as part of their fourth grade year, and so much more. More than 25 years later, those testaments speak for the sparks Betty created for her students.

What Were the Catalysts to Keep Things Going?

In 1996, Betty was honored by being named a Presidential Awardee for Excellence in Elementary Science Teaching for the state of Illinois. At the awards festivities in Washington, D.C., she randomly met a female representative from the National Science Foundation who mentioned a program that assigned educators to the Office of Polar Programs and current science grants in the Arctic and Antarctic. Little did Betty know that her life was about to change – again! In 1998, she was chosen to be part of the Teachers Experiencing Antarctica and the Arctic (TEA) Program, and she found herself bound for Antarctica with a global team of scientists from the United States and many other countries from around the world.

As part of the Cape Roberts Project, a geologic drilling initiative, Betty not only worked alongside scientists at the U.S. McMurdo Station, but was responsible for the education outreach component of the project. Doing the hands-on work of a scientist was an excellent way to learn how to communicate that science, and it also helped the scientists learn ways to communicate their work to a broader audience. Every educator in the TEA Program was a sponge, absorbing every bit of information and transferring it back to not only their own class, but to classrooms around the world. Each TEA teacher shared the lives and research of the men and women they worked with. Through blogs (then called journals), educators in the Arctic and Antarctic "talked" with students each day, describing in detail the process of science and the technology employed, which for Betty meant sharing the design behind the geologic drilling activity, as well as describing the sediment cores retrieved from the Ross Sea Basin (Trummel, 1998).

When Betty returned to her class seven weeks later, the excitement was contagious – both for herself and her students. She could clearly see that students were more engaged, more knowledgeable, and more excited about science, particularly polar science (Cape Roberts Project, n.d.).

This did not stop once Betty was home, as she spent countless days presenting to learners of all ages, passing along the information about the Cape Roberts Project and Earth's polar regions. At that time, little was published for primary school children and books for adults were also fairly limited to tales of the explorers of the early 1900s, the Heroic Age in Antarctic exploration. Audiences were eager to learn more about this remote place on our planet, preserved for science and exploration. With each presentation, Betty's confidence grew and she became a better speaker and was honing her skills of creating that sense of wonder about the polar regions.

Since very few people get to travel to and work on the continent of Antarctica or in the Arctic, Betty had become a member of a very small group that was dedicated to studying, protecting, and sharing these places with those who will never have the opportunity to experience the magic of being there in person. Life and teaching became a polar palooza!

When the next multi-national geologic drilling program was being organized, a call went out to teachers who would like to be involved. Betty was chosen to be one of six educators representing the four participating countries of the ANDRILL Project. The ANDRILL Research Immersion in Science Experience (ARISE) was comprised of three teachers from the United States and one each from New Zealand, Italy, and Germany. This is the program that forged a new connection between Betty and Matteo Cattadori, the Italian educator on the education outreach team. Working alongside scientists in Crary Lab at McMurdo Station, Antarctica, (Trummel, 2007) the ARISE educators participated in the research and shared their experiences with a wide range of audiences across the globe (ANDRILL Project Iceberg:: Blogs:: Betty, 2006). This was very similar to what Betty had done eight years earlier with the Cape Roberts Project (Pound et al., 2019).

One of the key deliverables of the ANDRILL ARISE program was a fabulous package called the "Flexhibit." Designed by ARISE educator Luann

Dahlman, and supported with multi-media materials such as photos and videos by ANDRILL's media specialist, Megan Berg, the Flexhibit became a very big part of Betty's classroom and outreach efforts after she returned from the ice. Created to become a flexible exhibit concept, the Flexhibit had five themes of science and research in Antarctica. The themes of "Antarctica Today," "Antarctica's Ice on the Move," "Diatoms," "Sediment Rock Cores," and "Antarctica in the Past, Present, and Future" – and the myriad of activities that came along with those themes – could be adjusted for use by multiple grade levels and presented in both formal and informal educational settings. All of it was designed to bring science alive for learners of all ages, and once students were taught the material and activities, *they* became the teachers (Huffman et al., 2021).

Thousands of children Betty taught and reached with this program were connected with real-time science research, and her students became confident speakers and presenters as they hosted Flexhibits year after year for the remainder of her teaching career. Betty stepped back and allowed the pupils to take the lead, and adults who attended the Flexhibit each year were astounded at how much her students knew about polar research and the science of this remote region. What proud moments there were, watching 9- and 10-year-old students share their knowledge and excitement with parents, administrators, peers, and everyone they came into contact with (Trummel and Dahlman, 2008).

The collection of books and printed materials, polar artifacts, posters, polar clothing/gear, and media such as photos and videos grew each year, all contributing to a lively and immersive experience in Betty's classroom. Literacy stations created to enhance student learning during the language arts block of teaching time often focused on science and social studies/geography topics. Every day was full of wonder and learning.

While all of this was happening, another great partnership was formed. Attending the 50th anniversary of the National Science Teachers Association convention in San Diego in 2001, Betty connected with a new colleague. Gunnar Jonsson from Lulea Technical University in Lulea, Sweden had attended one her presentations on her education outreach and science work in Antarctica. Lulea is just below the Arctic Circle, and Gunnar's work on environmental education, sustainability, and teaching pedagogy and content to future educators was a great fit for the work Betty had been doing. They decided to initiate a teacher exchange program between Lulea and Betty's school district in Illinois. At this time, Betty had also joined the ranks of university teaching as an adjunct professor, teaching "Science Methods for Elementary Educators." Not only were young students affected, but preservice teachers benefited as well.

For nearly 20 years, until the global COVID-19 pandemic paused the program, Gunnar and Betty spearheaded exchanges between their countries. Swedish teachers came to the United States, and U.S. educators traveled to northern Sweden and above the Arctic Circle to learn from their educational counterparts. This inspirational program was funded by participants themselves, with the exception of an initial grant to bring the first six Swedish teachers to America. Being in each other's classrooms, exposing students and teachers to rich cultural experiences, and sharing pedagogical practices was the highest form of professional development. Learning with and from each other in real-time field experiences and classroom situations was unparalleled for all of the teachers involved. This excitement and knowledge trickled down to the students in incredible ways (Trummel, 2012).

Two additional Antarctic experiences filled Betty with awe and motivation to keep sharing her work and the amazing polar regions with everyone she could. The WISSARD (Whillans Ice Stream Subglacial Access Research Drilling) Project in the 2012–2013 science field season in the McMurdo Sound region, and about 800 miles away on the McMurdo Ice Shelf, gave her new information and perspectives on science research based on hot water drilling compared to the geological drilling she had witnessed in both the Cape Roberts and ANDRILL projects. It was an opportunity to keep learning through the scientists and their research, and transfer knowledge to others through education outreach efforts while on the ice and well beyond her return to the United States.

The Homeward Bound Program in 2016 was an entirely different direction, though. Created by an Australian visionary expert in leadership training, Homeward Bound was designed to bring a group of 70–80 women in science from around the world together for a leadership training program which would culminate in a voyage to Antarctica. Think science, leadership, and a trip of a lifetime, sailing for 21 days from Ushuaia, Argentina to the Antarctic Peninsula. With many shore landings and opportunities to connect with the wonders of marine wildlife, glaciers, icebergs, science research stations, ancient calderas, spectacular mountains, and so much more, it was an unprecedented adventure which inspired Betty to keep learning and sharing, even though at that point she was retired from full-time classroom teaching. Retirement – just a word – did not matter, because her commitment to learning and education outreach was stronger than ever. Her own sense of wonder was on high alert!

Instead of learning alongside scientists and communicating their work, this was the optimum setting to focus on her own senses and being completely independent as a learner. Sailing south, she could witness gradual changes in the environment, and enjoy wildlife sightings and icebergs along the way. Once the ship reached the Antarctic Peninsula and outlying islands, shore landings were times to investigate and learn, to take in pristine land and seascapes. Traveling in inflatable zodiacs through a surreal polar seascape was magical; every sight and sound magnified, the cold air feeling both biting and exhilarating. Betty was surrounded by a panorama of icebergs and steep cliffs gleaming in the light. Massive glaciers spilled right down to the sea, the groaning and creaking of the glacial ice an indication that at any moment crevasses might give way and chunks of ice would topple into the sea. Mountain tops looked like they were covered with folds of whipped cream or slathered with vanilla icing, thick and smooth.

Moments to meander along the shore and stop to sit on rocks and piles of snow became time to observe gulls and terns gliding overhead. Bergy bits drifted by as penguins gallivanted in and out of the water; their quirky style and movement on land vastly different from their smooth gliding and turning about in the water. Seals surveyed the humans in their environment, keenly aware of these two-legged visitors. Seemingly awkward and lazy on land, the humans witnessed that seals are not so awkward in the water; their many unique adaptations enable adept movements in the ocean, key to survival.

Betty would bend down to investigate life in a tide pool, marveling at the tiny fish and crustaceans frantically darting about in the harsh Southern Ocean environment. She contemplated the importance of every single member of the vast and complex Southern Ocean food web and how changes in global climate are affecting this fragile web of life (Antarcticness – UCL Press, 2022).

The enveloping silence gave her time to pause and be immersed in the beauty and enormity of this pristine environment. It is a silence you can kind of feel. It seems primitive. In the absence of human sounds, it was time to look, learn, and rediscover her own sense of wonder. There were the natural sounds of cracking ice, glaciers rumbling, and seals barking. When you can hear your heartbeat, you feel the wildness. It is easy to say that Betty was mesmerized by what she witnessed on this voyage, as she had been with all of her polar experiences (Homeward Bound, 2023).

These encounters added to Betty's polar story and relationship with Antarctica. The compilation and blend of her individual and shared experiences had become part of the fabric of her life and career in education. Ultimately, these experiences were what she worked to convey to her students and friends through photographs, videos, and stories – her stories, that could inspire others to have their own adventures and discoveries in learning (Trummel, 2017).

Connecting to our planet through personal encounters is very powerful, yet so few people will experience this magic. It is critical to develop a sense of wonder for these places that need our protection. Awareness of our world cannot always be experienced first-hand, so it is an educator's responsibility to find ways to share this with learners (Trummel, 2016).

Case Study 2: Lessons for Life

Betty's colleague, Matteo Cattadori (Rovereto, Italy), and the creator of the Research and Education Svalbard Experience (RESEt) not only had a profound impact on his high school students, but on Betty as well. An incredible project, it still stands as one of the *best* examples of how educators can make a difference in the lives of their high school (or any level) students, and the long-lasting outcomes of such a project.

As previously mentioned, Matteo Cattadori and Betty met as part of the ANDRILL (Antarctic DRILLing Project) in 2006. Little did they know that years later, they would still be collaborating and working together on exciting educational projects. Through the years, they visited each other's classrooms and took part in professional development presentations in both the United States and Italy. Their involvement in Polar Educators International as part of the Council and Executive Committee continually reinforced their commitment to polar science education.

Exactly ten years after meeting, they managed to answer one question that has always puzzled every polar teacher or educator: "What might happen to a whole class of students when they are engaged in a real polar expedition?". This opportunity came with the RESEt (Research and Education Svalbard Experience; http://resetsvalbard.altervista.org/) a three-year long project aimed at planning, organizing, and funding a polar expedition with scientific-educational purposes.

The whole project was carried on with Matteo's class of the high school Liceo Fabio Filzi in Rovereto, Trentino, Italy. Consisting of 19 girls and two boys, RESEt was aimed at strengthening the initiative and independence of female students in order to let them be ready for post diploma and career decisions more focused on personal attitude and wishes than on social and cultural conditionings. It was an immersive science project, culminating in an expedition to the high Arctic Svalbard Archipelago region in July of 2016. Betty joined the expedition as an additional educator/chaperone and helped with English translations for the RESEt website.

Matteo's students raised €45,000 to help fund this expedition. Accomplishing this was a major feat in and of itself. They held bake sales, tutored other students, sought and obtained corporate sponsorship, sang on the streets of Rovereto, and even became live store mannequins to earn a percentage of the sales in certain stores. Students appeared in just about every venue possible and were also on the radio and in the news. It was a massive, all-out effort to gain the funding they needed to make this dream come true. In addition, funding was needed to make a documentary film about the RESEt expedition and their experiences in Svalbard (Cattadori et al., 2016).

Two years later, as the dream was realized and the plane approached the airport in Longyearbyen, in the Svalbard Archipelago, anticipation had built to a place that few students will ever truly understand. The students would realize a dream of two years in the making – to see a polar environment firsthand, and be immersed in learning in a way that is far outside of traditional classroom or textbook experiences. They had been taught more than subject matter. What they experienced had more to do with life, communication, and economic skills. Matteo had taught them determination, cooperation, and leadership, as well as how to follow your dreams. These skills were taught by modeling, encouragement, and thinking outside of the box. He had created a sense of wonder surrounding the polar regions, and his excitement and motivation transferred to his students in meaningful ways.

The week in Svalbard was not only full of learning, but was a showcase of students gathering information to teach others. Carefully crafted activities were divided among the students: taking photographs and videos, recording the sounds of Svalbard, taking part in scientific experiments and collecting data, learning more about the flora and fauna, creating watercolor paintings of the tundra biome, a radio interview broadcast back to Italy, and so much more! Betty watched as Matteo's students went far beyond the walls of any building – the world had become their ultimate classroom. What a proud moment in education! And, it was all documented by two filmmakers who followed every move of Matteo and his students. Incredible! What a way to record this amazing program to share with other teachers, students, and the general public – creating a sense of wonder not only for those on the trip, but to spread the beauty and fascination of the Arctic with a much larger audience.

As part of the time in Svalbard, there was a warm-up day hike to climb alongside a glacier, cross the top plateau, then don micro spikes to descend another glacier back to Longyearbyen. Next was a three-day backpacking adventure to be fully immersed in the region. What better way to fully comprehend such a magnificent environment than to walk its rocky tundra landscape, climb the gullies of fjords and skirt their plateaus, meander across enormous braided rivers, and slide down steep slopes that tumbled to the edge of the Greenland Sea? Everyone was in awe and enchanted with the Arctic land and seascapes.

From RESEt student/participant, Angelica Pergham:

How can I describe RESEt? The RESEt Project was far more than a school project. It was a life project.

Imagine to be 16 years old and try to raise money to do your dream travel, the project you have thought up and organized.

Imagine to go and knock on the door of many business managers and convince them that your idea and your project are worth the investment as a sponsor. Imagine to spend your free time doing activities to raise money and having extra lessons just because you want to know more and more about the topic.

Imagine to cooperate with your classmates, highlighting each other's talents, supporting each other in the difficult moments and rejoicing together for every achievement. Imagine to realize that you've made it and after two years of hard work, take a plane and reaching the North Polar region, just because you believed in it and put much effort in your work.

And after all that you're only 18. I think there is no theoretical lesson, no laboratory that could have given me the same experience, confidence and personal growth. Looking back at that now, six years later, I feel every day the sign of that project in the person I've become, in the choice I made and in the way I face up every challenge in my life.

What is striking are the long-lasting effects of a project such as RESEt. The 20 who embarked on the Svalbard trip have continued to meet with Matteo and Betty in the six years post-expedition, and all discuss the impact of these life experiences. Also revealing are the words of the parents of these students, as presented in a letter to Matteo upon the return of the group to Rovereto. Here are key translated excerpts from that letter.

Dear Prof. Matteo Cattadori,

To be honest, it felt like this was more than a project . . . a dream, and your enthusiasm and determination has now also spread to us. We knew our children wouldn't give up in front of any kind of difficulty. Now the journey has ended and we are awaiting your arrival, but, deep down, it's as if we were with you. In these days, thanks also to the website you have created, we were able to share your adventures, listening to the voices of our children, the dream achieved.

Thank you for giving the school and alternate route, capable of forming and not only educating our kids. This experience has served to make great, or better adults, this generation of young people. You taught them to cultivate the dreams, to understand that the commitment, enthusiasm, determination, courage, collaboration, and also the sacrifice and the fatigue are the ingredients to obtain, in life, great results. We would like to thank you with all our hearts.

For our children you are more than a professor, you are a teacher of life. You have been able to get them to love school. You made them pay attention to situations, become capable of designing and organizing events and activities, and above all, be able to think and reflect in addition to the materials and the school desk. You put them in contact with the world that is both geographical and anthropological, creating relationships with significant figures, thinking minds, people in the world of culture, of the market, science, politics; and opened routes which will become a path for their future.

Thanks to you our children were able to enjoy an amazing adventure, the emotions and images importuned on their minds will accompany them for life. They will have an indelible memory that they will be able to tell with pride, today to friends, tomorrow with nostalgia to their children.

Isn't that what we should strive for as educators: to create learning environments filled with excitement and wonder, and to encourage our students to follow their dreams and find rewarding learning experiences around every bend?

Conclusion

A sense of wonder cannot be forced; it must be cultivated and nurtured in ourselves and others over time. Of course, not every pupil will feel this, but they certainly will not if we do not make the attempt. Educators can provide this in a classroom through projects, presentations, or even time with family and friends. A teacher's enthusiasm is contagious and students feed off of the tone a teacher sets, whether daily or long term.

How educators present themselves as role models and mentors is key to creating classrooms that promote learning and discovery. Life-long learning is an essential component of that wonder. Paying it forward to others, helping them find this aspect of life and learning, has long been the goal of both Betty and Matteo. Their paths have been and continue to be shaped and sculpted by their science and specifically their polar experiences (2022).

References

ANDRILL Project Iceberg:: Blogs:: Betty. (2006) https://web.archive.org/web/201509 22115851/http://www.andrill.org/iceberg/blogs/betty/index.php.

Antarcticness – UCL Press. (2022) https://www.uclpress.co.uk/products/180739.

Cape Roberts Project | Antarctic Research Centre | Victoria University of Wellington. (no date) https://www.wgtn.ac.nz/antarctic/research/past-research-prog/cape-roberts-project.

Cattadori, M. et al. (2016) 'RESEt project.' http://resetsvalbard.it/.

Family Nature Summits. (2023) 'About us,' Family Nature Summits. https://family-naturesummits.org/about-us/.

'For Betty Trummel, M.S.Ed. '91, The world has always been her classroom.' (2022) https://www.myniu.com/article.html?aid=2009.

Homeward Bound. (2023) 'Homeward bound – STEMM women in leadership.' https://homewardboundprojects.com.au/.

Huffman, L. et al. (2021) 'ANDRILL's education and outreach programme 2005–2008: MIS and SMS project activities during the 4th IPY,' *Terra Antarctica Publication*, 15.

Kelly, L.B. (2018) 'Draw a scientist (science and children).' https://www.nsta.org/draw-scientist.

Pound, K.S. et al. (2019) 'ANDRILL ARISE: A model for team-based field research immersion for educators,' *Polar Record*, 55(4), 251–273. https://doi.org/10.1017/s0032247419000056.

Trummel, B. (1998) 'TEA: Tea_trummelfrontpage.' http://armadaproject.org/tea/tea_trummelfrontpage.html#calendar.

Trummel, B. (2007) 'Educators immersed in science research in Antarctica,' *Illinois Science Teachers Association Spectrum* [Preprint].

Trummel, B. (2012) 'International partnerships for professional development,' *Delta Kappa Gamma Bulletin: International Journal for Professional Educators*, 79(1).

Trummel, B. (2016) 'My earth science educator story,' *The International Geoscience Education Organization*, 2016.

Trummel, B. (2017) 'Science roadshow.' https://scienceroadshow.wordpress.com/.

Trummel, B. and Dahlman, L. (2008) 'The ANDRILL ARISE educational outreach program: Educators immersed in science research in Antarctica,' *International Earth Sciences Symposium* [Preprint].

Index

Note: Page numbers in *italics* indicate a figure and page numbers in **bold** indicate a table on the corresponding page.

For Product Safety Concerns and Information please contact our EU
representative GPSR@taylorandfrancis.com
Taylor & Francis Verlag GmbH, Kaufingerstraße 24, 80331 München, Germany